THE SPACE AGE NOW

A PERSONAL HISTORY OF MACHINES, TECHNOLOGY AND LIFE, FROM 1960 THROUGH 2020, NARRATED BY A SURVIVING CREW MEMBER FROM THE 20TH CENTURY

PAUL VLACHOS

NOMADCAT PRODUCTIONS LLC

"THE SPACE AGE NOW"

First Edition - 2020

NOMADCAT PRODUCTIONS LLC

www.paulvlachos.com

ISBN: 978-0-9914535-0-4 (Paperback)

ISBN: 978-0-9914535-1-1 (EPUB)

For My Nuclear and Extended Family

For Santo and Elko and every one else I have met along the road

Special Thanks to Cordy Ryman and Tonya Audyn Stiles, who helped me with ideas, language, and editing

A human being survives by his ability to forget. Memory is always ready to blot out the bad and retain only the good.

— Varlam Shalamov

1

THE SPACE GUN

When I was five years old, I had a small metal pistol. I had many toy guns, but this one was for outer space. It was made from two pieces of stamped metal, had a working trigger and was painted blue. When you squeezed the trigger a few times, a little gadget would start to spin inside and create sparks, which you could see through red plastic panels on the gun housing. At the same time, it made a loud whirring noise. Suddenly, you were transported to a distant planet, doing battle with galactic foes. With my steel toy Colt M1911 pistol, I might be in some World War II battle. With the Peacemaker, which had a cylinder that popped out and could take fake bullets, I might be in some remote canyon. The space gun took me to the galaxies, though, and that was where all of my friends were also going at the time.

I would play space games with my best friend, who lived across the street. First, we would blast off and leave the planet. We would sit on the ground somewhere, usually with our backs against a wall, and get ready for liftoff. Then would come the most important part, the countdown. We would scream in unison, counting down from ten to zero together. We would make "vroom" and rumbling noises for a long time, blowing air through our closed lips, after which we would be in space. From that point onward, everything that happened, usually in one of our backyards, happened in space. I had another friend who did not go out much. I would visit him a few blocks away and we would play "Lost in Space" or "Star Trek," visiting distant planets and dealing with strange villains. It was the mid-1960s and we were little kids.

2

ROOTS

I WAS BORN IN 1960 AND GREW UP IN YONKERS, NEW York, on a dead-end street at the top of a hill. Our neighborhood was called "Mohican Heights" and all of the streets had Native American names. My whole world centered around that block, the local public school, and the five-block radius around them. My father commuted to work every morning on the New York Central Railroad, soon to become the Penn Central, later to become Metro-North. He would leave the house and walk down the hill every morning, holding his briefcase and wearing a homburg, then stride through two miles of tree-lined streets to Tuckahoe Station. He would catch a train with other men of the suburbs, all on their way to earn a living for their families. On the train, he read the New York Times to find out what was happening in the world. The Times cost ten cents an issue then if you bought it at the newsstand, less if you subscribed for home delivery.

My mother stayed home. She was happy to be a homemaker and a mother. She loved to read mysteries. She was a fanatical knitter and made all of us sweaters, hats and scarves. She cooked three meals a day, some of it modern, mid-century cuisine, but also some elaborate Greek dishes that she learned from her mother or Jewish, Russian, Italian and Irish dishes from the women she had grown up with in the Bronx.

She loved to spend as much time as possible with me and my brother. She was always encouraging us to write, draw and sing. My mom and dad took their responsibilities as parents and as adults seriously and they worked hard to make ends meet. They were busy all the time. My dad worked long hours and, when he was home, he would be making home repairs or taking care of the yard. My mom maintained a tight social network with her brother, her parents, and her friends. When she was not working or on the phone, she was in the middle of the block, talking with a neighbor. She was a connected woman.

My grandparents were all born in tiny, remote mountain villages in northern Greece at the end of the nineteenth century. They had no cars, no running water, no electricity, and no telephones. They came here by steamship as children. My father was born in New York City in 1922 and my mother grew up in the Bronx, although her mother had returned to Greece for my mom's birth. She traveled by ship. Lindbergh would not fly solo across the Atlantic for another two years. My parents grew up in tenements, but they had electricity and lived in New York City at a time when new technology was exploding all around them - the first half of the twentieth century. Automobiles, electricity, telephones, telegraphs, radio, motion pictures, and x-rays were all young technologies then and they were drastically changing the way people lived.

My parents got married and moved to the suburbs in their early 20s, in 1948. By this time, they had seen people fly across the Atlantic for the first time, had witnessed the birth of commercial aviation, and had lived through a world war with submarines, jet planes, radar, and atomic bombs. They had witnessed the birth of television, the commercialization of frozen food, the invention of "miracle drug" antibiotics, electric typewriters, long-distance phone calls, and the ubiquity of home appliances. They settled in a quiet suburb, just north of the Bronx, and began to raise a family. My brother was born in 1951 and then my mother became ill with tuberculosis. I was born in Manhattan nine years later and brought home to Yonkers, New York.

3

THE BIG HAND

My friend, Roland, and I grew up and played in the dead-end. It was a quiet street with tall trees growing in the yards. Every autumn, leaves would fall from the pin oaks. We would rake the leaves into big piles, then spend the day jumping into them. It was not odd, before plastic lawn and leaf bags appeared on the scene, for people to burn piles of leaves. In winter, snow would fall and settle up to our waists. We dug tunnels through the snow. Roland's back yard had a cherry tree that we used to climb. My family had an apple tree. In the summers, my dad put an oscillating sprinkler on the grass and we would jump through it over and over.

When someone put a sprinkler on their lawn, the water would puddle into the street. We would disconnect a hose and leave it at the top of the block, then trace the water down the street with twigs, leading it along in rivulets and tiny puddles until it began to head down the street on its own. We called this "The Atomic River." We were obsessed with the globs of tar on the street in front of the Glynn's house. They had dripped down from a paving machine one prehistoric day before we were born and remained in big splotches there for the entirety of our lives. On hot summer days, the tar would bubble and we loved to pop the bubbles. Sometimes, water would leak out of them. The only tool needed was a finger or a small twig. Soon, we were burning leaves with a magnifying glass and making a lot of noise with a Big Bang cannon.

Our house was near the top of the dead end. There was one more house on each side of the street before the road ended. Roland's house was across the street from my house. Every house on the block had a back yard and most of them were surrounded with low cyclone fences, which we quickly learned to scale. Everybody hung their clothes out to dry in the warm weather. We had a clothesline on a pulley that was bolted onto a corner of the house and then to a tree. Roland's family had a rotating outdoor clothes rack that was like a square tree. We loved to spin it around.

Our milk was delivered every day by a milkman, who drove up the block in a small truck that had no seat. He wore a uniform and drove standing up. The truck had hand controls on the steering wheel. We had a Styrofoam-lined zinc milk box outside the front door. He would walk up with a few bottles in his metal hand basket, leave the milk, take the empties, and then drive off. Our mailman walked up the block with a heavy sack of mail on his shoulder. My mom would give him a can of beer on hot days. A few telephone poles brought the power and the phone lines up the block. They had spikes driven into them so that linemen could climb up.

In the actual dead end, someone usually parked a car on the left-hand side, but there was a tiny ecosystem on the right side – dirt, a few trees and some bushes. Dominating this was a huge hunk of granite, as big as a refrigerator on its side.

Predictably, we called it "the rock" and would often start a plan by saying "I'll meet you at the rock." You could cut to the block on the other side of the dead end and the next neighborhood if you walked through one of the yards at the top of the block. This was how I went to school every day.

Everybody knew each other. The families did not socialize much, but the kids all played together. The ties among the kids were different than the ties among the parents. They were two different social networks with some overlapping ties. We also knew kids from the adjoining blocks, but did not see them as often.

When we were four or five years old, we would run to the dead end whenever the big, yellow city garbage truck arrived. The street was too narrow for them to turn around, so they would back up the hill. One guy would get out, then empty the garbage cans into the truck as the other guy slowly steered it down the hill in first gear. After every other house, the sanitation workers – we called them "garbage men" – would activate the hydraulic compressor. A large steel claw emerged from inside the truck with a lot of noise, then scooped and pulled the garbage further inside. We were thrilled to watch this mechanical show and never missed it if we were outside. We called it "the big hand."

4

THE HOME PHONE

Our house was a classic, small Cape Cod, built after the war for returning vets and baby boom families. World War II had ended just 15 years before I was born and many of the dads in my neighborhood were vigorous combat vets in their 40s. At the time, people were leaving the cities and urban blight for the suburbs. My block was sleepy and nobody parked in the street, so there was plenty of room to play.

All non-postal communications with our house revolved around the ivory-colored, Bell System rotary-dial phone mounted on the wall in the kitchen. We would one day call it "the kitchen phone," after we added another phone. Nobody used the term "home phone" yet. It would have been redundant. This phone had an extra-long, coiled handset cord so that you could walk around, do work, or go through a door to the enclosed porch for some privacy while talking. This white cord would grow yellow and dirty with age, but it was remarkably tough. I learned at a young age how to disentangle a tangled handset cord. You held the cord and let the handset dangle and slowly twirl until it sorted itself out.

The handset had tiny holes in the ear and mouth pieces. It would get dirty from constant handling. Wiping it down and cleaning it once in a while was a normal part of life. If the cover pieces got dirty, they could be screwed off and washed, at which point you would glimpse the technology inside – a small speaker at the top of the handset and a microphone on the bottom. The phone had a loud ringer. This could be adjusted with a small metal lever on the bottom, but it usually remained at the same volume - medium - and it could not be turned off. The mute button did not exist yet.

A yellow version of the same phone hung on the wall of the dining area in my grandparents' house. They were my mom's parents and lived in another part of Yonkers. Important numbers were written on the wall in pencil. There was no speed dial and telephones had no way to memorize numbers. My grandparents were old, so the wall was handier than pulling out the address book. This phone, the Western Electric 554A, was a reliable electro-mechanical device and it produced high quality sound with its little speaker.

The model was derived from the Model 500, which remains a jewel of twentieth-century industrial design. It felt indestructible and had plastic edges that did not cut, perfect bevels, and a handset that fit well between one's shoulder and one's ear. It was the phone I grew up with. We had it for a few decades. The idea of a phone model changing often would have been odd. The signal fed out from the house over a twisted pair of copper wires. These led to a trunk line in the street that led to a switching house somewhere in Yonkers. All signals traveled over copper wire. They were not broadcast through the air.

Much of daily life revolved around this device with the rotary dial and the long, coiled handset cord. My mom would spend a lot of time each day on the phone, having conversations with friends and family. When she was young, in the 1930s, telephones were not used for frivolous, long personal conversations. Phones were used for business. She discovered great social value in it and spent hours on the phone, usually while cooking or cleaning, talking with friends and gossiping. It let her stay in touch with people. The phone may have been invented with one thing in mind but, eventually, it became something else.

When the phone rang, it meant the outside world was calling. Somebody would run to pick it up and this created random interactions: "Let me see if he's here," "Hold on, I'll get him," and "How are you doing?" were part of the daily dialogue. "It's for you" was a common phrase, as the handset was being handed over to somebody else. Nobody was associated with an individual phone line unless that person lived alone. You called a household and asked to speak with someone by name. You hoped that the right person would answer so that you did not have to make small talk. Telephony at the time encouraged you to be polite to people you might not necessarily care for. Other people around you could also hear you talking.

Long distance and international calls, which usually had to go through a special operator, were big events. We did not use area codes locally in the 1960s and even into the 1990s. You had to remember only seven digits. Every house had two telephone books: a "white pages," for residential numbers, and a "yellow pages," for business. You made sure to have the latest edition. New editions were published every year. You might put the older one in another room or use it to prop something up, but it was important to always have a current copy. If you needed more help, you could pick up the phone, dial 411, get an actual human being on the first or second ring, and ask for help finding a number. It was called "Directory Assistance" and it was exactly what it sounded like. If you needed help placing a call, you would dial "0" and a live operator would pick up and help you.

5

MID-CENTURY CONSCIOUSNESS

My mom would often start the day with a call from her brother, who was an advertising executive and owned his own agency from the 1960s through the 1980s. He and my mother talked every morning for years and I would overhear him running new TV ads by her. It made me feel important to think they were discussing an idea that would end up on television. Over the course of the day, she might call her mother, her friend who lived in Bayside, Queens, or one of the relatives who still lived in the Bronx. They would talk about other people, the news, and low-level gossip. It was hard to fathom what she was discussing. I was just a kid and would tune her conversations out. Life in the suburbs was more isolated then. It may have been true in the city, as well. The telephone was a lifeline.

You walked around in a bubble, without communications. You might check in from a pay phone on the street. Gripping the handset, tethered to a phone, you would ask what you had to ask, say what you had to say and scream what you had to scream. Then you would move back into the real world of consciousness and external stimuli, free from electronic intrusion, moving through the universe with only the thoughts in your head. There was no constant contact with other people nor the expectation of it. We were ships on the open sea with no radios.

There were no answering machines. If one appeared in a spy movie, it was only to suggest high technology. The idea of leaving a voice message was not on the radar. If you had something important to say and you could not reach somebody, you had to wait until you spoke with them. If someone else answered the phone, you might leave a message – "Tell him to call me. It's important." Usually, you waited to connect and, when you finally spoke with them, you might preface the conversation with something like "I've been trying to get to you for a while."

When I was a young child, I did not make any calls. I watched other people talking on the phone. Somebody might hand me the phone, which I would approach as though it were the communal totem that it was. "Come and take the phone, Paul." It would be handed to me and I would take it, then talk into it while others watched and listened, smiling or frowning. Many conversations were shared affairs in this way. Privacy was not so common then, nor was it expected.

Since many conversations were public, you might speak cryptically to conceal what you were saying from those standing around. Not until a few years later, when my father added a telephone upstairs, was there a semblance of privacy. It was still the same phone line, though. If you were having a personal conversation on the phone upstairs, somebody could quietly pick up the phone downstairs and listen to it. My brother and I would always be listening for a small click in the background that would signify someone was eavesdropping. There was a protocol, of course, if you picked up the phone inadvertently and discovered that it was in use. You would say something like "Oh, sorry." or "Oops." Privacy was cherished, but not always available.

Later on, the one person I did call was my friend Roland, who lived across the street. Of course, we tried to communicate in other ways, being young and living with our bedroom windows 100 yards apart, across the street from each other. We had the primitive flashlights of the era, powered by weak D-cell batteries, with dim, yellowish incandescent lightbulbs, but we were too young and lazy to learn Morse code. We did the constant "dot dot dot, dash dash dash, dot dot dot," but that was it. We thought about stringing a wire across the street and somehow using that with tin cans. We did not have money for any electronic devices, nor did we know of any that existed for us. We eventually got some cheap walkie-talkies, but they were useless. There were no other options. We did not have much to discuss, but we could talk all day.

When together, we would play our favorite game: "let's pretend." Whole universes were conjured up with words and imagination. Another zero-technology game we loved was hide-and-seek. On the phone, we might speak about what was going on in that very moment or what had happened earlier in the day. We might discuss what was on television the previous night. We all knew what was on TV because everybody watched the same 7 channels.

6

PHONE PROTOCOL

ACROSS THE STREET, IN ROLAND'S HOUSE, A TELEPHONE sat on a small desk in the front entry hall. A chair sat under the desk and there was a lamp on the table. It was a formal phone station. They also had an older phone in the basement. I now know that it was a Bell 302, a beautiful, functional art deco sculpture of a device, designed by the firm of Henry Dreyfuss, with thick, non-coiled wires connecting the handset to the body and the body to the wall.

It was heavy, a remnant from the dawn of the Plastic Age. The sound quality on a call was fantastic. You had to exert some force with your finger to move the metal rotary dial. This dial would wind back with a satisfying grind after each rotation. Printed in the circle in the center of this dial was the phone number. From Roland's house, I would call home, at 779-4260. That was my first number. My mom taught it to me with the original telephonic mnemonic device of the twentieth century, the exchange: "SPencer9-4260." This phone number was probably the first thing I was taught to memorize. My mom also taught me the alphabet, which she wrote for me in chalk on a small blackboard when I was 3 years old. She spent a lot of time with me, reading, talking, singing, schooling me in the ways of communicating. We spoke and listened to each other all day.

Phones were stationary. You needed to sit in one place when you spoke on the phone. Some people installed extension cords and walked around, holding the receiver in one hand and the phone in the other. Others bought the extra-long, coiled handset cables. There was plenty of curling up in the corner with the phone and a pillow, having long conversations. Conversations may have lasted longer when we had only a home phone. Did people talk about different things then? Possibly.

The actual phone conversations may or may not have been more relaxed, but they were certainly less frenetic and distracted. There was almost no multitasking while on the phone. Nobody had speaker phones. Call-waiting did not exist for us. When you were on the phone and someone was trying to call, they would get a busy signal. "DON'T TIE UP THE PHONE" was a common admonition. It would diminish you a little bit. Your call was less important than whatever call was expected. It was also a way to nag you. "Get off the phone. I'm waiting for a call. Do your homework."

My mother had a phone in her house growing up. My father did not. This was common in the late 1920s. They were both from New York City, but they met in the Catskills, in a remote hotel called the Star House. One phone line ran through that isolated valley upstate. It was a party line and they both had vivid memories of how it worked. The phone rang a certain number of times to identify which household the call was for. You picked it up if it was for your house. If you wanted to eavesdrop on your neighbors, all you had to do was to pick up the phone quietly and listen. The dearth of lines and the primitive switching of the day dictated this. Any phone was better than nothing, though. The country was still being wired for electricity and telephony in the first half of the twentieth century. You could hear the actual network. When you picked up the receiver and got a dial tone, you knew that you were entering another dimension.

7

THE BASEMENT

ASIDE FROM LIGHT BULBS, THERE WAS ALMOST NOTHING electronic in the basement of that house in Yonkers. It was a dark room down a steep flight of stairs. There was a light switch at the top of the stairs. My father eventually added one at the bottom, but the house was built with a switch only at the top. That basement terrified me. I was young and had a vivid imagination. The light switches in that house were the same as almost every other one I have ever used – a tiny lever that went up or down. Lamps had their own push/pull or twist switches.

Muscle memory would kick in when you reached over and twisted a switch. You automatically executed however many turns it needed to switch on or off. Occasionally, you would run into a wall switch that was illuminated, but most switches for the first few decades of my life did not vary. The design was simple and effective. Some switches did not click and were utterly silent. My dad explained that they were made with mercury inside. This would close or open the circuit when you flicked the switch. On the rare occasions when I encountered a switch with a dimmer, it was something I would play with for a while. Dimmers were cool.

My father put night lights everywhere, probably for practical purposes. He was a practical man. We had them in every house we occupied. A few of these nightlights had small incandescent bulbs, like clear Christmas lights. He also found some that glowed with a tiny orange light and seemed to never wear out. They were not LEDs. They put out very little light, but they seemed to make him happy.

At the bottom of the basement stairs, you made a sharp left and walked across the room and through some cheap, wooden louver doors. Another left turn took you to the oil furnace, an electro-mechanical device with no transistors, no tubes, and certainly no microchips. When that house was built, microchips did not exist. Commercial transistors were in their infancy. The furnace seemed like a sentient being, going on and off at the whim of the thermostat. The thermostat was on a wall upstairs at an adult's eye level and my father would stand there and adjust it with his glasses pushed on top of his head so he could read the tiny numbers.

Next to the furnace were a washing machine and dryer, and across the room was my father's workbench. He owned only hand tools at that time. I liked working there with him, and he bought me a set of kid's play tools, but they were no fun after I turned five or six years old. The saw didn't cut, the nails were tiny, and the pliers were made from cheap, soft metal. Real tools, on the other hand, fascinated me. I would go down there and play with them all, from the vise grips to the hand saws to the calipers that he had milled in engineering school. I loved playing with the tools and would use them on scraps of wood and metal. Nothing else in that room used electricity except for the washer and the drier, which were hallmarks of post-war technology and middle-class success.

Above my father's tool bench was a flashlight, powered by two D batteries, with a small, incandescent bulb. It was better than what had come before, twenty or thirty years earlier, but not by much. The beam was dim and yellow. I loved to play with flashlights. The first one that I owned was in the shape of a ladybug, painted red and black, with a bulb at the tip of the head. It was all metal, had a sliding metal on/off switch, and was powered by two AA batteries. It fit in the palm of my hand, which made it even more fun to play with. The main thing I did with it, and all other flashlights, was to lie in bed or on the floor in a dark room and shine it on the ceiling, sometimes whipping it around, fast, from side to side. It was entertainment enough to shine a light in a dark space, to feel the magic of flipping a switch and creating a beam of light.

In the main part of the basement, which my dad had finished in classic 1960s faux wood paneling, there were a couple of old chairs, a lumpy couch, a lamp, and a shelf filled with some books and board games, including "Risk," "Monopoly" "Battleship," "Twister" and a chess set. Most toys then were not electronic. We had a dark mahogany upright piano from the late nineteenth century they had bought for my brother. It sat next to his drum set. There was a dark storage space under the stairs that you could reach through a low door. Inside were some old beach chairs, a big wicker picnic basket from Greece, and a few toys. There was a closet near the door to the garage. This is where my father stored the cases of Pabst Blue Ribbon that he bought at the beverage wholesaler in downtown Yonkers.

The only sophisticated device in the whole lower level of the house, through a heavy steel door at the bottom of the stairs, was the car. This sat in the sunken garage, under the master bedroom, down the steep slope of the driveway. Shoveling that driveway in the winter was tough. You had to heave snow up and over a tall wall. My dad never bought a snow blower or a power mower in that house. Summers were spent pushing a heavy rotary mower across the lawn and being jealous of the neighbors, who all owned gasoline mowers. We would sharpen the blades of that push mower with a hand file. The sound of those two-stroke gasoline motors on the block was the sound of progress, the future.

8

THE ROCKET 88

THE CAR WAS A BLACK 1960 OLDSMOBILE ROCKET 88. IT had a V-8 engine with a manual shifter that sat on the steering column, also known as a "three on the tree." This meant it had a three-speed gearbox and was not a stick shift. It had no computers or microchips, the only electronics being the alternator, distributor, lighting and wiring harness. None of this was new technology. Most of it was 40 years old at the time.

They bought the Oldsmobile the year I was born and it was the first car we owned that had a radio, according to the only living witness, my brother, who is ten years older than I. Our father - an engineer, a minimalist, and a child of the Great Depression - had finally broken down and ordered an AM radio with that car, which also came from the factory without seatbelts. My father ordered a set of lap belts from Sears and installed them in the driveway one Saturday afternoon in 1965, when I was five years old. I remember watching him bolt them into the floor of that car and then testing them to see how they worked. I don't remember using them, though.

Our next car was a 1968 Oldsmobile Cutlass and my brother swears that our father had to special order it with no chrome ornamentation of any kind. It also had an AM radio and you would save stations on it by pulling out one of the buttons and pushing it back in again. It was a simple mechanism and it worked well. The car doors locked with buttons that you pushed down to lock and pulled up to open. Windows opened by hand with cranks that you turned. There were no key fobs. Our garage door also opened by hand, although electric garage doors did exist by then. Roland's house had one for their double-wide garage door. You needed to get out of the car and use a key to operate it, but I was still impressed.

My father's brother, Uncle George, was more of a libertine than my dad. He would buy a new Buick Riviera every two or three years and he would order every option. His car had the first power windows and door locks that I ever experienced. All the doors had ashtrays and cigarette lighters built into the arm rests. On his dashboard, he stuck a compass, a globe with directional markings on it, floating in liquid. I loved to watch that ball spin every time we turned a corner.

When we drove to the Catskills or Cape Cod on vacation, my brother and I would curl up in the back seat with pillows, barely awake, at the barbaric hour of 5am, when our father would always leave to beat the traffic upstate. There never was any traffic. Our dad, whom we loved dearly, was a driven man in a race against his own demons. Those demons always woke up before he did until he got much older. We could not read in the back seat because it was a dark and bumpy ride. All we could do was look out the window at life passing by. We would try to nap. We played word games, like "Ghost," and we talked. We sang songs together, "The Bear Went over the Mountain" was a big one, and we sat there until we arrived. I would look for one particular roadside attraction just past Kingston, New York – a cannon that belched a five-foot flame every thirty seconds.

We rented a farmhouse in a sleepy valley that was filled with extinct dairy farms. The last farm had stopped operating 20 years earlier. An abandoned creamery sat on the other side of the creek and I loved to explore the ruins, which were filled with tubing and pieces of pumping gear. My grandmother used to walk with me in the freezing waters of the mountain-fed creek and we would pick mint and watercress. Next to a small bridge nearby was a big yellow earth grader. It was parked there permanently and smelled of ancient heavy grease and oil. This huge mass of steel impressed me before I even knew what it could do.

The farmhouse had a back room that was filled with boxes of stuff from when it was a working farm and they used to feed dozens of people at a sitting. There were cartons of china, curios, and tools. Nothing came from the modern age. There were no wires, electrical appliances or any objects made of plastic. Walking through that door was stepping into a time capsule from the 1920s.

My dad always carried a toolbox in his cars. It was spartan, but useful, based on hard-won experience: a few hand tools, some bolts, some wire and a bulb or two, a flashlight and a pair of gloves. Money was tight, so he used to change his own oil. He would park the car at the top of the driveway slope so he could get under it easily. Automotive technology was simpler then and you could work on your own car if you were willing to get dirty. Everything was mechanical.

The garage was narrow, but we kept some boxes along the walls. Rakes, shovels, the push lawnmower and the hedge clipper, essentially a gigantic scissors, hung on the wall from hooks. My bike was there, along with two galvanized steel garbage cans. I had an old wooden milk crate in that garage. It was filled with metal objects – a faucet, pipe sections, old batteries, hinges, a rusted railroad spike, some bolts and other prized junk that I found. This stuff formed the basis for many games. The garage door was made of wood and split into four hinged sections so it could fold when you raised or lowered it.

Mine may have been the last generation before the plastic Big Wheel tricycle. My first tricycle was an all-metal machine from the 1950s they had bought for my brother. My parents had enough faith that they would have another kid, despite my mom's bout with tuberculosis, that they saved everything. They kept all of his bicycles for me. The tricycle was the first, and my first bicycle was another beautiful machine from the 1950s, with chrome fenders, wide handlebars, a kickstand and a red steel rack, screwed securely onto the rear. It was a single speed. My next bike was my brother's second bike, what people called a "newsboy bike." It was bigger and heavier than the usual bicycle, had a single speed, and was almost indestructible. It carried double wire baskets in the rear and the handlebars were as wide as steer horns. Two chrome rods curved from the handlebar stem to the fork with an enameled steel badge bolted to them.

9

THE GROUND FLOOR

ON THE MAIN FLOOR, THE KITCHEN WAS THE NERVE center of the house. I spent childhood days crawling on the kitchen's off-white linoleum, patterned with small pebble shapes in a terrazzo pattern. Off the kitchen was a small dining area with an archway into the living room, where a big window faced the street. Opposite this archway was a door that opened to an unheated, enclosed porch with jalousie windows. This room felt like you were entering a boy scout cabin in the Adirondacks, with varnished maple furniture and cushions that had autumn leaf patterns. There was one overhead light on a switch.

The living room and dining area were carpeted in a brown wall-to-wall rug with a bumpy pattern on it. There was a couch and a table, a few lamps and a wooden cabinet. On the other side of the house were a small den and a small bathroom. This den was my bedroom until I moved upstairs. Next to it was the master bedroom, which also had a window on to the street. This bedroom was over the garage. It had red carpeting and a bed that I would hide under.

In the kitchen, the phone hung on the wall next to the sink. The refrigerator door had a long, heavy chrome handle that looked like a car hood ornament. It was placed vertically, along with a smaller matching one, inverted, on the freezer door. Their shape hinted at the Jet Age. You would pull the handle forward to open the door. Inside, it was standard issue white enamel and plastic with rubber-coated wire shelves. The freezer was frost-free. The sink had a disposal for churning up waste food. My parents told me at an early age how dangerous this device was. You would put the stainless-steel stopper in the sink, turn on the water, twist the handle in the stopper and a terrible grinding would rise up from beneath the sink. I would conjure up a mental image of my hand being ground to a bloody pulp in a gigantic pencil sharpener.

It was a small kitchen and there were few appliances aside from the stove and fridge. We had a blender from the early 1960s and a toaster from 1948, the year my parents were married. It had great sentimental value for my parents and my father nursed it along for decades with the occasional home repair. Our can opener in the 1960s was crank-operated. It swung out on a hinge from the wall and took a few turns with the crank to open a can.

My mom's technological revolution was in the kitchen as much as anywhere else. She had grown up in a household where all meals were made from fresh local ingredients. This was the late 1920s and the 1930s. There was no frozen food and they owned no electric kitchen appliances when she was a little girl. Refrigerators were expensive. Most families had iceboxes, which were cooled by a giant block of ice that was delivered every few days by the ice man.

In her life, she saw the advent of frozen food, prepared foods, electric mixers, blenders and electric percolators. When she was an adult, my mom used her pressure cooker and electric frying pan all the time. These were modern conveniences and labor savers. She favored Wesson corn oil instead of the olive oil she had grown up with. These things were sold to her as the future, along with microwaves and frozen foods that were cooked in their own plastic pouch.

My dad was so poor during the Great Depression that he and his brother ate mayonnaise sandwiches – no meat, no cheese, just mayo - that their father would bring home from work. He was another convert to modern conveniences, although he would sometimes deride them as "gadgets" or "gizmos." I grew up with orange juice made from the A&P house brand orange juice concentrate or whatever was on sale that week. We ate a lot of canned and frozen foods, as well. It was all about convenience and cost and the basic food groups, or my mom's version of them.

There was no radio in the kitchen, but we had a small tabletop AM radio in the dining area, three feet away. There was little space in either room. The significant electronic item in the dining area was a small black and white Panasonic television, which appeared there in the mid-1960s. It sat on a small buffet table with a wood-grain Formica top, the kind of table that stood in a little rack of four, ready for use at parties and other events. The small TV sat on that table permanently. I watched the first moon landing on this TV on a July evening in 1969.

As with all cathode ray tube televisions, you would turn a switch and then wait for it to "warm up." The screen would stay black for 30 seconds, sometimes longer. If you looked into the guts through the rear vents, you would see an orange glow, then the screen would come to life, along with the sound. The on/off switch and the volume knob were one and the same, a tiny knurled knob with a small, but satisfying click to turn the TV on or off. The channel knob was not illuminated, the numbers were printed on and it clicked hard every time you turned the channel. If you wanted to watch UHF stations, you pushed and turned a small inner knob. This tuned in channels 48, out of Connecticut, or Channel 21, out of Long Island. There was a single speaker behind a steel grill on the front of the set and the tube was about thirteen inches, measured diagonally. There were no lights on it to indicate anything, not even power, whether it was on or off. There was simply the tube, which would start to glow in black and white.

A single, pivoting antenna telescoped from the top and a circular UHF one folded up from the back. You did not need to move the antenna much to catch Channels 2, 4, 5 and 7. Sometimes you had to fiddle to get Channels 9 and 11. The signals all beamed up to Yonkers from the top of the Empire State Building. Since we were at the top of a hill fourteen miles north, the signal strength was pretty good.

The television picture quality was awful, but we didn't know it, as we could only compare it to our neighbors, who all suffered from the same 480 scanning lines. It wasn't the fuzziness or low definition that bothered us, it was the reception. Even with a good signal, you often had to tune and tease and tweak the controls to get a stable picture. Sometimes, local atmospherics would change, and you had to move the antenna. We had a big antenna on the roof, connected with a flat wire that was almost half an inch wide and ran into the house through a hole in the wall. I don't recall this being connected to anything. All I remember is constantly adjusting the TV-mounted antenna.

Anything could interfere with the television signal. Running the vacuum cleaner or a blender could mess up your picture. Sometimes, you would get bars running vertically on the screen, fat bars like a wide t-shirt stripe, slowly or quickly moving up and down and taking the picture with it, frame by frame. For this, you would grab the tiny "Vertical Hold" knob on the back of the set and slowly turn it until the bars stopped, taming the picture back into the frame. The black and white scanning lines danced with electrons and we had TV. Sometimes, we watched while we ate. That television lasted a long time. The faux wood grain on the metal casing became all scratched up and dented with age.

Roland's family had a color TV, which was more of a rarity then. The big set faced their main dining table. They had a separate dining room, but ate most of their meals in the kitchen, watching that big television. It was an MGA, an early Mistubishi. I was impressed by color television, but my parents had no interest in it and would not buy one until the 1980s.

10

OTHER TELEVISIONS

THERE WERE TWO OTHER TELEVISIONS IN OUR HOUSE. The main set was in the den, the small room off my parents' bedroom. When I was young and that was my first bedroom, the TV was in their room, on a wooden stand with round casters. They moved it into the den when I moved upstairs. The other TV was in my brother's room, which was also upstairs. Up the carpeted stairway, his room was to the right and mine was to the left. The two rooms were separated by a bathroom and a tiny hallway. There was no attic in that house, so the upstairs bedrooms had slanted ceilings.

The television in the den was a black and white RCA with a nineteen-inch tube. As with the other TVs, it had manual controls, all knobs, but there was a small incandescent bulb behind the dial which would dimly light up the channel numbers. Compared to modern TVs, the picture quality was awful – fuzzy black and white – but it was good for its time. I spent my earliest days viewing TV from the floor with two pillows under my head. When they let me, I would watch the same lineup of shows every night for an hour or two. My mom always told me not to watch in the dark or I would ruin my eyes. I watched in the dark, anyway. My eyes were ruined, but it was probably from all the reading I did. She also told me not to stomp my feet when I got angry or I would develop flat feet. I stomped them anyway. One day, the pediatrician told me my arches were almost flat. My mom sat there with a grim look on her face. I'm convinced she was trying to establish a medical condition for getting me exempted from the draft, which was in force at that time for the Vietnam War. I was only six or seven, but she was thinking ahead.

I would watch TV for an hour or two at night before they made me stop: Truth or Consequences, Petticoat Junction, Combat!, Batman, Bewitched, Time Tunnel, Star Trek and other shows. Saturday mornings were for cartoons. I enjoyed books, as well, so I somehow struck a reasonable balance between television and reading. One of my earliest memories is of being three years old and lying on the den floor, pillows under my head, silently watching JFK's funeral procession, the riderless horse with boots backwards in the stirrups and Walter Cronkite looking sadder than a grown man should look.

The den was small. When my parents watched television in there, my dad would sit in a wing chair and my mom would lie on the couch with some knitting. They would focus on the TV. I don't recall watching a lot with them. They watched shows that I was not interested in. When we went on vacation to the Catskills, there was no TV. When we went to Cape Cod, we stayed in a small, rented efficiency apartment. There was one TV set, so we had to watch together. In my early teens, I sat with them in Cape Cod and watched Nixon resign, live on television.

My brother bought the TV that was in his bedroom. He was 9 years older than I and had saved all his earnings for a year to buy it. It was a nineteen-inch black and white unit and it sat on a brass-colored wire rack. He could have bought a wired remote so that he could lie in bed and switch the channels without moving, but our father talked him out of it. My brother would still rue this and mention it over fifty years later.

He watched the New York Jets, whose games were blacked out in New York City and re-broadcast later in the evening. He would invite me into his room and lecture me on the greatness of Joe Namath and the other Jets. The New York Yankees were awful then, but I was a Yankee fan and watched them anyway. My father had been a lifelong New York Giants baseball fan, but he lost interest when the team moved to California in 1958. He didn't watch much baseball until later in life. He listened to news radio, though, and he liked to watch the news and movies on television. His generation grew up watching everything at the movie theater, including newsreels. When he got older, it was perfectly natural for him to go to the movies by himself.

My mom watched some strange stuff during the day, but no soap operas. She watched game shows and old movies while she was ironing, which she did in the living room. She would set up the ironing board so that she could see the small TV in the dining area from an angle. More often than not, she would listen to me as I read out loud from a book while she worked. I was too old for Sesame Street, which debuted when I was nine. By that point, I was also done with morning cartoons and was devoting my time to weekly shows, such as Gilligan's Island and Hogan's Heroes, or old movies. Roland and I watched Creature Features and Chiller Theater, two programs that showed only horror movies. We loved the classics from the 1930s and would watch them at the relatively late hour of 8:00 pm. Then, we would call each other surreptitiously to discuss Frankenstein, Dracula or the Wolfman. The discussions would often spill over into the next day.

I loved war movies, and this was when Hollywood, twenty years after World War II, was recreating and mythologizing the American experience in that war. I lived for those movies. My suspension of disbelief was strong. My cinematic aesthetics were weak. At that age, I could watch those war movies and get lost in the heroics. My other memory of war was at the dinner table, when we would watch short clips on the evening news of the fighting in Vietnam. It was nothing like what I saw in the movies. They would show napalm strikes in black and white on patches of jungle, small groups of soldiers talking into a field radio, occasional combat, and interviews with tense-looking officers. I would finish my dinner, then run to the other TV set and put on Hogan's Heroes. My brother, who was in his late teens, watched the news from Vietnam with more interest, especially after the government began to broadcast draft lotteries as special events on live TV.

11

VACUUM TUBES AND DEPRESSION BABIES

In the living room was my father's stereo system, something I did not touch. Soon after I was born, we began to move from working poor to the middle class. At that point, my dad purchased a modest Scott amplifier. It was powered by vacuum tubes and was built to amplify the music, color the tone, and direct it to speakers. He bought a companion Scott tuner, again made with vacuum tubes, for the radio signal. His DUAL turntable plugged into the amplifier and that's how he played music. People called this kind of setup a "stereo" or a "stereo system."

In the 1950s and early 1960s, many families had a standalone console, usually a long, low piece of wooden furniture that would and integrate the stereo gear. You unfolded the top on hinges to expose a turntable and a few control knobs. With the console units, records sat on shelves between the speakers, usually behind thin sliding doors made from particle board. These units were a link to the tremendous radio sets that Depression-era families kept in their living rooms during the golden age of radio. The consoles were more than enough for our grandparents, who worked hard and enjoyed music, but only on special occasions.

They had seen the birth of radio broadcasting and lived much of their lives with radio as their great connection to the world. Commercial television did not take off until the 1950s. The technology was there before the war, but it was experimental and there was no real programming. Few people owned a television set immediately after World War II. Families would sit together and listen to the radio – news, sports, serial programs, drama and comedy. My parents spent the first 25 years of their lives with radio and records. This was the technology of The Great Depression, which had formed them.

My grandparents still had their ancient radio consoles. They still functioned, so why bother to replace them? My father's mother, who lived in a tenement on Dyckman Street, in Washington Heights, had a tremendous radio console, a fine mahogany unit almost the size of a refrigerator. It had a huge knob in the middle and the whole design followed from that. It took forever to warm up. A new set of tubes would have helped. The reception was terrible, but it had a ton of push-buttons, which made it fun to play with when I was a kid. She had a smaller record player and never moved past 78s. 78s were heavy records that rotated at 78 rotations per minute, or "RPM." They were scratchy, the sound quality was not great, and they had one song per side, but it was the standard format for years and people were happy with them until the next format came along.

My mother's parents bought a console stereo record player and that's the technology they kept until they died. It played LPs, or "Long Playing" records, as well as 78s. I have fond memories of wild Greek dances blaring out of that unit during family get-togethers, with people spontaneously breaking out into solo dancing in the small living room. They also had a piano in the living room, a low upright piano, with an old metronome sitting on top.

The metronome was made of dark hardwood and looked like an elongated pyramid. You could unhook a tiny latch, remove the front panel, and wind a small key on the side to put tension on the spring. You gently pulled the pendulum from its captive slot and it would start to tick. It was slow and majestic and made a solemn "tick tock." You adjusted the speed with a sliding weight on the pendulum. I loved playing with this mechanical gadget, which was a product of eighteenth-century technology, but was still unsurpassed by anything else at the time. My paternal grandmother had a music box on her bureau that would play a little song when you opened it. This also wound up with a spring mechanism. You could turn it over and watch the rotating drum with the little nibs on it hit the reeds as it turned.

My dad was a jazz buff and my mom, who worked at Decca Records in the late 1940s, would smuggle him in to recording sessions. A highlight of his life was getting to sit inside the studio during a Billie Holliday recording session. He had acquired a large collection of 78s by the time he was forty and these remained in a wooden cabinet that he built. That cabinet moved from house to house, always in the basements, along with a small phonograph dedicated to playing 78 RPM records. 78s needed a bigger needle and many of the newer turntables were eliminating the speed as an option. These records became a fixture in my life. By the late 1980s, he finally bit the bullet and threw them out. Before he did that, though, he recorded dozens of them onto cassettes so that he could still enjoy the music.

My dad bought separate components instead of a console. When he assembled the living room stereo, which remained his default system until the end of his life, he installed it in an old piece of furniture. It was a dark, polished wooden cabinet. The doors snapped shut with little magnets. He mounted the two Scott components, amplifier and tuner, on the right side. He ran cables through a hole inside and put the minimalist, ancient DUAL turntable on the left side, on a slide that pulled out. He kept a few records underneath, but most of them went on separate shelves in the living room. When closed, it looked like a low china cabinet but, with a few hand movements, it transformed into a stereo system. He had good "AR" brand speakers and, when you played a record on that system, it filled the room with music.

I grew up with jazz, show tunes and classical music downstairs and my brother's rock music upstairs. Record covers were on the floor, propped against the cabinet while they played, so they became familiar, totemic objects for me. I would stare for hours at the covers of the "La Dolce Vita" soundtrack, Louis Armstrong records and Glenn Gould's Mozart sonatas. At some point, my dad bought a pair of heavy, gray Koss headphones and a long extension cable so he could get some peace and privacy. He would put on the headphones, which were a specialty item for audiophiles back then, lie on the floor and disappear while life went on around him. I later took those headphones and they carried me through many years of surreptitious listening in bed.

He was proud of that system and kept it up to specifications. I would go with him to the hardware store and watch him use the tube tester. He would pull a suspect tube and take it to the store, pop it in the machine and watch what the meter said. If the tube was weak, he would replace it. Stereos needed more maintenance back then. You had to keep an eye on the phonograph needle to make sure it didn't get too dull. You had to wipe down records to remove the dust before you played them. This record cleaning would eventually achieve fetishistic proportions by the 1970s. I owned nothing at the time, so life was simple. There was always music in our house, especially on the weekends.

The radio was on a lot, usually news radio, which I grew to hate at an early age. Even then, before my consciousness was fully formed, news radio was an ugly, grating squawk in my ears. I had no need for the news, weather, traffic or sports. It was just compressed blare. My mom would listen to pop music and sing along as it played. I would toy with the car radio when I was in the front seat and always stopped at the pop, rock and salsa stations. I liked to explore the airwaves.

We owned another radio, stowed away in a closet, ancient but still functional. It was "portable" in the late 1940s, but was as big as a large toolbox and it brought in shortwave signals. In 1974, after we moved back to Yonkers, I used that radio to listen to a broadcast from far away of Muhammad Ali taking down George Foreman in Zaire. Most of our other radios were modern transistor units, but this one ran on vacuum tubes. Transistors were the miracle technology of the 1960s in many ways. They had been around for years, but in specialized applications, such as hearing aids. With the MOSFET transistor, first deployed in 1960, you had a device that needed little power and could be scaled smaller in size as the technology progressed. These transistors are the building blocks of microprocessors, digital and analog circuits, and all digital electronics. They would not appear in my life for years and, when they did, I was unaware of it at first.

12

MODERN AMENITIES

ASIDE FROM MY FATHER'S STEREO AND A FEW LAMPS, there was nothing else in the living room that required electricity. We had a windup clock that nobody wound. This was technology from the previous century. This was the biggest room in the house, but we used it the least. We watched TV in the den and that was more of a communal than a personal activity. My parents had grown up clustering around their console radios, so it was not a reach for all of us to gather around the TV set.

I had a compact phonograph that I inherited from my brother. It was the size of a manual typewriter case and the top unhinged to reveal a metal platter and two knobs. It was most definitely low fidelity, although the term "lo-fi" was not used at the time. It had a powerful motor that was useful for playing games. You could spin objects by placing them on the felt-covered platter. You could hold your finger on the platter while it spun, and you could scrape your finger on the needle to produce amplified scratchy noises. There were no input or output jacks.

There were no jacks on the next one I inherited from my brother, either, when he went off to college. That next phonograph could play LPs – 33 1/3 RPM records. It was also a record changer and you could stack a few LPs at once for continuous play. When the first record finished playing, the tone arm would put itself back into its holder. The next record would then drop down the spindle onto the platter. The tone arm would raise itself off its post, jerk to the left, and lower itself into the groove at the edge of the record. It was a foolproof system and you could count on it for a few hours of uninterrupted sound if you stacked up enough records. If you bought a new turntable or record player, one of the specifications in the literature would be the number of records you could stack on the spindle.

I liked to watch record changers in action. The crudest spindle mechanisms had an offset area near the top. A thin piece of steel would stick out and hold the records up with the help of a stabilizer arm. My father's turntable had a more sophisticated mechanism. Three small pieces of steel popped out of the spindle shaft, so there was no need for the stabilizer arm. The record would sit there evenly. The three pieces would disappear simultaneously into the spindle and the record would drop. I would sit for hours to watch it repeat this maneuver. You could stack only one record at a time on this turntable.

I look back with wonder at how few devices we owned or needed to own. Nothing in our lives needed a microchip. We had a couple of transistor radios, but that was the extent of miniaturization in our lives until the 1970s. The phones were hard wired with a twisted pair of copper wires. These wires went to the phone network from a utility pole across the street. Everything was physical except for the radio and television signals, which were magical and amorphous, but even they seemed to possess tangible properties. When we moved to another house in Yonkers a few years later, we had a ghost image of the World Trade Center towers on the television screen due to the transmission logistics.

My mom had a Singer sewing machine in a polished mahogany cabinet that dated from the mid-1950s. It ran on an alternating current motor, but the basic apparatus was technology from the nineteenth century. We also had her Royal manual typewriter from the early 1930s, which she would use to type a letter or fill in a form. My brother got a manual Olivetti typewriter by the time he went to college. The typewriter and sewing machine were old, reliable technologies. Even electronic watches were exotic in the 1960s. All of our watches wound up. I had no real need to wear a watch at the time.

The last thing to describe on the main floor of that house was a device that connected it with the upper floor – an intercom. My brother and I cannot come up with a good reason why our parents ran an intercom from the kitchen to his room. It was a small house and my mother's voice was strong enough to carry. The intercom was a compact beige unit from Sears that my father installed on a weekend day in the mid 1960s. He ran a wire from the main unit, in the kitchen, to the sister unit, next to my brother's desk. He had to drill a hole through the ceiling to do this.

When they wanted to reach my brother, they would press a button in the kitchen and a loud buzz would sound in his room. Either party could press the button and speak, then wait for the other person to respond. My dad did not like raised voices. I don't know whether they inflamed his sensibilities or simply reminded him of the tenements he had lived in.

I cannot think of another good reason for that intercom. We always felt it was a strange purchase for a guy who was so frugal with his money.

More useful was the phone jack that he installed at the top of the stairs, just outside the bathroom. It connected to the main phone line and it was a big deal at the time. The phone company frowned on anybody installing an extra jack. They did not want anyone touching their network. He did it himself, and this impressed me. He installed one of the old four-pin jacks and suddenly we had a phone upstairs. It changed my phone habits. You could now have some privacy while you were on the phone. Anybody else in the house could still pick up another phone and listen in, but at least my brother or I could curl up at the top of the stairs with a pillow and settle into a conversation without having to whisper.

13

TOYS AND TAPE RECORDERS

There was no television in my room upstairs. One day, a clock radio appeared. It was a horrible unit, made of beige and white plastic, with an analog clock and a useless radio. I recall no joy or success listening to it. I didn't use it for the alarm function, either. I don't know why they gave it to me, but maybe it was from some Depression-era notion that I ought to have a radio. More likely, they bought it for the dining nook and discovered that it was useless. Rather than toss it, they put it in my room.

At some point in the mid-1960s, my father installed a small air conditioner in my window. It was an inefficient unit that made a great, howling noise. It rumbled and blew out semi-cool air. It ran all summer and gave me an early appreciation for white noise. We had a ritual every summer of hauling all four air conditioning units up from the basement and installing them in windows around the house. We would take them down again in the fall. We would also bend the fig tree over in the back yard every fall, cover it with tar paper and plastic, then shovel dirt on top so it could survive the winter. My dad then had to go to my mom's parents, his in-laws, and repeat these same chores for them.

My brother remembers, before I was born, only one air conditioner in the house, in my parents' bedroom. He says they used to watch TV on hot nights in their bedroom, then he would go upstairs to his room, which was oppressively hot, with no air conditioner. My parents would turn theirs off and go to bed. Our childhoods were slightly different, as we were 9 years apart. Technology was changing. My father was making a little more money by the time I came along. Still, we had more in common than not.

I inherited my brother's American Flyer train set, which does qualify as an electric device. It was a glorious toy. I would lay the track down, then hook up the transformer and run it for hours. It came with an eyedropper bottle of a chemical called "liquid smoke." You put a few drops in the engine's smokestack and it would puff smoke. I loved that chemical smell. I also had a set of tiny slot cars and loved to stage actual crashes with them. Another toy I inherited from my brother was an "Erector Set," by Gilbert, which contained a small alternating current motor. You could connect this motor to gears and wheels to make them spin with a grinding noise. It contained steel girders that bolted together with small nuts and bolts. These came in a round cardboard container that had a metal top that pressed shut. The set came with its own tools, as well.

From my brother also came an "Electric Football Game," a sheet metal board painted to look like a football field. You would set up the individual players. The quarterback was a figure that did not even pretend to look human. He was a little gadget with a spring-coiled catapult arm that could "throw" a tiny felt pill when you released it. You would set up the opposing teams and then turn the game on with an electronic rocker switch. The board would vibrate madly with a sound like a gigantic electric razor. The players would move in all directions on the board until you turned it off. Even at that young age, when I was hugely forgiving of discrepancies with reality, it was hard to fantasize that this was connected to football. It was more like a "let's play with electricity" experiment.

The other electronic toy that came from my brother was a helicopter. There was a vertical pole, then a cross pole at the top of it. The helicopter sat on one end of this cross pole and a counterweight sat on the other. A control panel was wired to the unit and you would start the chopper's rotors, which raised it into the air. Another switch would tilt it forward and you could then fly it in a 3-foot diameter circle for as long as you liked. I played with this a lot.

There were a few other electronic toys available to my generation, but I had none of them. Radio-controlled cars existed, and the game "Operation" always enticed me with its television commercials, but I did not own them. I liked to play with the vacuum cleaner and enjoyed the feeling of power it gave when you threw the on/off switch. We had an upright vacuum. It was not self-propelled, but it had a headlight on the base. My grandparents had an ancient, steel canister machine with a steel on/off switch that you pushed with your foot, like the high-beam switch on an older car.

I would also power up my mom's portable hair dryer and use the tube to blow hot air in all directions. If a toy glowed in the dark, it was especially cool. I had a Creepy Crawlers set, which contained metal molds. It came with a small hot plate and bottles of "Plasti-Goop," which you would pour into the molds, then cook in the hot plate. I shoplifted a bottle of glow-in-the-dark Plasti-Goop from the local toy store, called "Tim and Tom's," on Central Avenue. When they were cooked, we would hold them up to a light bulb, turn the lights off and watch them glow.

Every suburban basement then had a stack of board games sitting on a shelf, along with a few decks of cards. You would pull them out on rainy days or when there was a gathering. These games had more in common with nineteenth century parlor games than anything else. They were familiar to most people and were low-tech ways to while away the time. Board games, along with chess and checkers, were engrossing and called for some degree of interaction and were a minor escape from reality.

My dad gave me a Swiss stopwatch at around this time. It was wind-up, analog and a precision instrument. I played with it endlessly. I would watch it run, stop it, reset it and do it again. I might incorporate it into some game that Roland and I were playing. It would suddenly become a handheld instrument for exploring space or time. I also had a tiny Timex watch with a stainless stretch band. It had glow-in-the-dark numbers and hands. Anything glow-in-the-dark was a hit and we got a lot of mileage out of holding a watch against a lightbulb, then staring at it in the dark. The glow did not last for long.

My brother's room held two things that gave me a glimpse into the future. First, he had his own television set, sitting on the wire rack at the foot of his bed. He would close his door and watch television at night. I could see the white light coming from under his door. This struck me as the height of personal entertainment. Watching TV alone in a room was something I could look forward to when I got older.

The other object in his room was stashed on the bottom shelf of a bookcase, low enough for me to reach, and it was a tiny porthole into my own future. This was a Webcor "Compact Deluxe" reel-to-reel tape recorder. It was about the same size as my phonograph – roughly fifteen inches square and seven inches high. There was a plastic handle riveted to the front of the case so you could carry it when it was closed. The case was covered with a gray/black pebble finish. The cover had two latches on it and detached completely. At the back of the case was a space for the power cord and a small, square wired microphone.

Instructions were printed on the inside of the cover. You would push the empty plastic take-up reel on to the right-hand spindle, then thread a reel of tape onto the left spindle and pull it across the heads. You would catch the tape leader in the take-up reel and turn the reel with your hand until the tape caught. You spun it a few times to get more tape onto the reel. This took about thirty seconds. The controls were simple. A small switch turned on the power and a knob adjusted the volume. There was a multi-function bar, curved so your thumb could push it in. This made the tape move. You had to pull it out with some force to stop it. You moved this switch to the left or the right if you wanted to rewind or fast forward. When you wanted to record, you would simultaneously push a small red switch with your left hand and the main bar with your right hand. This would engage the record head. You could change the recording speed, as well, with a push/pull switch. The tape recorder felt futuristic to a seven-year-old.

Why did we have this machine? It could not record music in high fidelity. Home recording was not big then, anyway. 33 1/3 RPM records sounded better than any compact reel-to-reel. They were more convenient. My father probably bought the machine to rehearse business presentations he had to make. Plug in the square black plastic mic, hit "record" and speak into it. Then, listen back to what you recorded and ponder the ancient inscrutabilities of your own voice. I found other uses for this magical device.

It took little to keep me entertained at that age. A discarded kitchen faucet and some scrap metal would become a spaceship control room or the inside of an army tank. A small shovel could be a steam shovel. I had a vivid imagination and disappeared easily into a fantasy world. This tape machine was as high tech as it got, so I pulled it off the shelf in my brother's room all the time. Running the tapes at high speed felt like I was operating a rocket ship or a time machine. Touching the plastic reels while they were fast-forwarding felt dangerous.

It was great fun to listen to a recorded tape while it was moving fast or slow, in forward or reverse. It felt as though I were performing an electronic instrument. I began to record strange noises and tried to record conversations in my house, although the Webcor was too big to be stealthy. I would record music from a television, radio or phonograph by holding the microphone up to speakers. I never got bored with this tape recorder. I would turn it on with no reels and power the spindles backwards and forwards while I let them flap my fingers.

The tape boxes were interesting, too, from the cheap Shamrock brand to the more professional 3M and Scotch reels. My brain was young enough to absorb everything. The tape boxes had blank lines printed on the back so you could label them. I would lie there and study these boxes, reading the notes people had written on the back. The ability to record sounds was a revelation. It changed the way I thought about the world. If you could record light with a camera and sound with a tape recorder, you could collect things and check them out later. They were memory machines and the idea was planted in me that time and space were not only recordable, but also malleable.

14

PHOTOGRAPHS

There is one more device I should mention from the 1960s and that first house. It was a Kodak "Starlet" Brownie-type camera that I inherited from my brother. It took 127 roll film and had a very simple mechanism – a film advance lever and a single speed shutter. The aperture and focus were both fixed. All you had to do was advance the film and snap the shutter. It was essentially a point-and-shoot. The photo industry was trying to encourage people to buy film and pay for processing. There was a flashbulb attachment you could screw on to it, but I could not afford flashbulbs or batteries.

I took few photos at first. I did not need film to play with that camera, though. I would look through the viewfinder as though it were a live video feed. I would advance the non-existent film with the lever and then snap the shutter. I would repeat this over and over. I would open it up and marvel at the flat black innards and the empty film take-up spool. I would walk around with it hanging from my neck on a strap, like a junior paparazzo. Somebody finally gave me film and I took some photos, but most of these are lost to posterity. Those that remain are strange, black and white documents from someone who is four feet tall with no sense of visual aesthetics and a pre-adolescent sense of humor. There is a photo of me in the den from this time. Around my neck is the camera, next to me is the primordial phonograph and, in my mouth, for some odd reason, is an unlit cigarette. Did I think this was the what an adult looked like? Was I imitating someone I had seen? Did my brother pose me? I don't remember.

Cameras and snapshot photography were intertwined with family life. My mom documented everything. This came from a purely sentimental urge. She wanted memories – tangible, physical memories - so she photographed everyday life and special events. Later in life, it became a rote thing. She would pull the camera from a drawer whenever the birthday candles were lit up or the Christmas gifts were opened.

The Kodak Instamatic camera was a glorious tool for someone like her. When these cameras came along, they took the technical work out of photography. The film was a cartridge that you popped in - no muss, no fuss. I recall the satisfaction of ripping open the heavy foil packaging on the film cartridges and getting a whiff of the Kodak chemical stew from new film. You would click a button on the side of the camera, let the back swing open lengthwise and pop the cartridge in. Then you would snap it back together, push the advance lever a few times, and watch the black-on-yellow arrows go by in the tiny frame window. Eventually the number "1" would appear, after four or five thumb pushes, and you were ready to shoot.

Indoors, you needed a flash for that slow film, so you would pop a flash cube onto the camera. This was another time-saving device. The flash cube encapsulated four flash bulbs into one small cube rotated after each shot. My mom would pull a camera from the drawer and the shooting would begin. Sometimes she said "smile" and sometimes she didn't, grabbing a candid of somebody in a compromising or unsuspecting pose. I don't think there was a conscious philosophy behind her shooting, but there was a method. She wanted to capture moments from her life the way she was experiencing them.

When she was done with a cartridge, she would send it off, usually in some discount mailer envelope, or she would bring it to the local drugstore. This was 10 years before "15 Cent Photo" commercials, from a company in New Jersey that processed film for 15 cents a photo. Most of these photo labs did not specify glossy or satin prints, edgeless or border. Some would send back prints with serrated edges and the date stamped on the bottom border.

The final product of all this was the print. These usually stayed in a neat pile in the envelope. The prints came back in the mail and we would look at them. The small pile was handed around and we enjoyed them together. The photos might sit on a side table for a while or we might look at them after dinner. Sometimes, they were passed around one at a time, in a rotation, and they would end up in a pile again. They would go back in the envelope and my mom would put them away.

Some people let the mailing envelopes pile up in a box. My mom's final goal was to create photo albums. She would put the small packet of negatives in a shoebox and then arrange the prints into albums. She had grown up with old-fashioned albums. These were binders with heavy paper pages. You would paste "photo corners" – small adhesive right angles –

onto the paper. Or you might put one on each corner of the photo, lick them all and then stick them down. You could arrange the photos on a page however you wanted.

By the time I came along, she was using plastic sheets with clear pockets for each photo. You had to buy sheets the size of your prints. She would put the photos in the plastic sheets, then put these pages into a large loose-leaf binder that my father had brought home from work. These albums ended up on a shelf in the closet. This system took hold when I was around five years old and remained constant for decades.

Before that, my father took the bulk of the family photos and his aesthetic was different. He took snapshots, but they were more composed. He also took carefully framed landscape shots. My dad had a brick of a camera, an ancient Argus 35 millimeter, in a brown leather case, the kind of case that screwed into the camera's tripod socket. It unsnapped in two pieces and was called an "Ever-Ready case." Every control on that camera was manual and you needed to know what you were doing, especially with the slow film of the time. He took hundreds of slides, beginning with my brother's birth. The slides ended up, meticulously labeled, in metal file boxes that were about twice as big as a school lunch box.

He used an off-brand projector. Each slide sat in a thin metal sleeve and these went into long metal holders. It was a tedious system. When we had company, he would haul the projector out, load a slide box into it, turn off the lights and the images would glow, one by one, on a fold-up screen. People sat in the dark room and watched while they smoked, talked and drank. This did not happen often, but I have special memories of these communal viewing parties. He had put too much work into this system to even think about buying the Kodak Carousel projectors when they came out.

At some point, he stopped shooting slides and switched to film. At the time, film was seen as an improvement over slides for home shooters. Among pros and artists, it may have been different. He surrendered to convenience and bought a Canon C-35, a compact, semi-automatic camera with a good lens. Sometimes, he carried it on business trips abroad. He kept shooting photos, but not as often as my mom did. Years later, she had to abandon the Instamatic system in favor of a drop-in 35mm camera. The film loaded easily, but I think she missed those cartridges.

When I was 12, they gave me a Kodak Instamatic 110 camera. It took smaller film cartridges than the previous Instamatics, was pocket-sized, and my first camera that was not a hand-me-down. My brother had an Olympus Pen camera that shot half-frames on 35mm film. You could shoot up to 72 smaller photos instead of 36 standard ones. I also had a camera that my Uncle George had brought back from Germany after the war. He had bartered for it and gave it to me when I was kid. It was an ancient piece of technology that unfolded with a leather bellows and a shutter mechanism on the lens. It felt like an antique in the 1960s, just 15 years after he got it.

15

1960S SUBURBIA

THIS WAS THE HOUSE I WAS BORN INTO AND WHERE I spent the first 10 years of my life. I played outside all the time, even in winter. I dug big holes in the backyard and made mud with a garden hose. I would stir the mud with a big stick. I would climb trees and run through the neighboring backyards. Roland and I stole Civil War military markers from the local graveyard. We shoplifted toys from the local toy store and vandalized meaningless objects. We rode our bikes all over the neighborhood. I read books and I watched TV. I went to school.

In my room, I could hear my brother, across the hall, blasting records from the Stones, the Beatles, Bob Dylan and the Animals. I also heard the sound of his typewriter. We were all heavy readers and I was surrounded by books. Every room had bookshelves. I would browse them all day. I also went to the public library, where I wandered the stacks or looked through the card catalog for what I needed. I learned about the Dewey Decimal System and how to find a book. I would thumb through the long card drawers with the worn index cards, look at the different titles, then go find a book where the card said it would be.

I had a library card with my number embossed on a small piece of metal that was folded into the cardboard. The librarians would make an impression from the metal when I checked books out. They would stamp the due date with a rubber stamp into a card in the back of the book. It was an elegant, simple system that worked well. It was at the checkout desk in this library, in 1968 or 1969, that my father, who had just read an article in the paper, held his index finger and thumb about an inch apart and said "Someday, you'll be able to read a book on something the size of a cube this big." I am guessing he had read about microchips or computer memory. I was incredulous, but I believed him. Anything seemed possible at the time.

At my elementary school, P.S. 28, nothing was electronic. We had wooden desks and small steel and wooden chairs that had a shelf underneath for books. Dark green cards with white lettering were tacked high up on the walls, showing the letters of the alphabet in cursive. Teaching was done with blackboards and chalk. Our one experience with advanced technology came in the fourth grade. The school received an audiovisual kit and we made film strips by drawing with wax pencils on clear, matte 35 millimeter film with sprocket holes. The finished product was projected onto the wall for a fairly unsatisfactory experience. I'm sure it was an experiment in education that never caught on.

Gimmicks like that aside, my education mainly consisted of teachers and textbooks. I carried the heavy textbooks to school and back in an old leather satchel. School backpacks were not common at the time. The school year began with the purchase of pencils, erasers, glue, scissors, a loose-leaf binder and adhesive reinforcements. We presented current events with clippings cut from the newspaper. We had no television presentations in the classroom. It was a small school. Consumer videotape did not exist. At the end of the day, I would trudge seven blocks home, briefcase in hand. I would get to the final dead end, cut through the my neighbor's backyard, and be on my own block again.

At this house in Yonkers, my media consumption increased as I got older. I made an electric cave from one of the two closets in my room. It was a small space with a low door and, since it was at the top of the house, one of the walls was sharply angled. Some of the roof nails stuck through, sharp ends exposed, from the shingles that were just an inch on the other side of the wall. It was small, but the plywood gave it a warm feeling. I was the only one who could fit inside of it so, one day, I just moved in and made it a den of my own.

I put some pillows in there, along with a book or two. On the wall was a bare incandescent bulb in a ceramic socket with a pull-chain switch. All I had to do was find a splitter socket. This was a double light socket that screwed into the existing one. The lightbulb went into one socket. Into the other I screwed an adapter with an electrical outlet. Into this outlet went a three-way plug and into that plug went an extension cord. I scrounged these pieces downstairs from my father's workshop. Once I had light and power, the game was on.

I would bring the thirteen-inch black and white TV set in there and watch TV shows until my parents caught me. I brought radios in there and listened to them. I listened to everything that was on the New York City AM spectrum in the 1960s. I brought the phonograph in there and played records. The closet was too small for my model train set, but I hauled in everything else that could be plugged in and turned on. It was my first workshop. That phase in life did not last long. I grew too large for the space and we moved away soon after that, but there would be other crawlspaces in my future.

Maybe this was normal behavior, Maybe I was introverted. Maybe this proclivity came from my Greek mountain genes. My people were descended from a tribe in the Balkans, the Vlachs, who tended sheep. Each winter, they would decamp to the warmer areas, near Volos, leaving one man behind to tend the sheep. I imagine him sitting in a small, cave-like room, introverted and crazy with revelations, waiting for his people to return.

16

PRINCETON, NEW JERSEY

In 1970, my father's job took us to Princeton, New Jersey. I had to leave P.S. 28, in Yonkers, where I had gone from kindergarten through the fifth grade, to attend a private school in Princeton. We left the dead-end street where I had grown up, my best friend, all the local kids, my school, and the five-block radius that was my whole world. I have since met people who moved every year as kids and learned to deal with it, but I had a hard time. I was becoming an adolescent and Princeton is where I lost the innocence of childhood and an adult consciousness began to creep in. I took long bike rides by myself. I read books all day, lying in a hammock and drinking gallons of sweet iced tea made from a powdered mix. I collected and read thousands of comic books during what would later be called the "Bronze Age" of comic books. I learned how to be alone.

The house in Princeton was bigger than the small Cape Cod in Yonkers. It was a split level and had a bigger yard, front and back. We had the same car, the 1968 Oldsmobile Cutlass with no trim, no air conditioning, and no clock, but with an AM radio. We brought the same TVs with us from Yonkers and I inherited the thirteen-inch black and white TV from the old dining area. This went into my room and became my window onto the world of old movies.

Every night, I would put a towel under my bedroom door to block the light from my TV. I would plug the pointy white earphone with the eighth-inch jack into the side of the TV set, find a station and watch old movies for hours. I watched anything that came on, but I did begin to scan the local TV listings, just to be on the lookout for movies I might enjoy. UHF channel 48 was a favorite. I was used to watching color films in black and white. Occasionally, silent films played, as well as early "talkies," which is what they called films in the late 1920s that had a synchronized soundtrack.

My bedroom also contained a desk my dad made with milk crates that we pilfered from behind the local supermarket. They were the long ones, made from steel with plastic bottoms. We stacked two on each side, then placed a closet door on top and covered it with white Formica. I inherited the lamp on that desk from my brother. It was a plastic replica of a flintlock pistol, facing upwards. The plastic trigger was the lamp switch. The oval lampshade had a drawing of Daniel Boone on it, in full uniform and carrying a rifle. My brother and I stuck random little stickers onto that lamp shade over the years, mainly from plastic car models.

The telephones in this house were wired. Cordless phones and answering machines were still in the future. Mail came through a door slot, as opposed to the old, outdoor wall-mounted box in Yonkers, and my father finally bought a power drill and a power jigsaw. He also bought our first gasoline-powered lawn mower, a Toro, and I liked to steer it around the yard as I cut the grass. We bought a simple electric can opener for this house, but it never worked as well as the manual one from the old house. I spent hours at the Princeton University bookstore, sitting in the aisles and reading. This is where I first began to buy LPs.

We had all the books and the same radios from the house in Yonkers. Very little had changed in terms of technology. We did have to get a new phone number. The local radio stations were different. My brother went away to college that year and bought a slightly better stereo. It had two speakers that separated from the main unit to allow for stereo separation. The turntable folded down from the case and the whole thing felt more modern. I inherited his previous record player. My father still had his Scott stereo system. My mom had the same 1950s Singer sewing machine. Color television was still just a dream for us.

I hated my new school and felt different, suddenly ethnic, and ostracized. This may have had more to do with me than anything else, but I had come from a public school, with working and middle-class kids, to a school that had a lot of wealthy kids. It's not easy being a new kid in any situation, especially at that age, but I felt profoundly alone. I would not have survived that time without my bicycle, books, comic books and old movies.

In addition to the living room, kitchen, bedrooms, garage and basement, this house had a large room that we called the "family room." My first house had "the den," which was much smaller than this room and which had doubled as my dad's office. Most suburban houses then had what was variously described as either a "rec room," for "recreation room," a "TV room," or a "family room." This space usually held a TV or an "entertainment center," which was a combination console that had a stereo and TV combined. There was usually a couch and some easy chairs, along with a card table. If you were lucky, you had a recliner.

In most of the family rooms I saw, including ours, all of these furnishings were a bit worn and tattered. Most houses had living rooms that were off-limits, or the chairs and couch might be covered in plastic. The family room was where everything happened. It was the communal center. If someone was in there, watching a show on television, you had to either watch what they were watching, negotiate with them, or fight to change the channel.

This house had a crawlspace under the family room. As I had done with my bedroom closet on Iroquois Road, I sought out privacy in this crawlspace. Maybe it's a testament to the desperation of adolescence that I went here. I hated dark basements, but I went in there and discovered a large pile of bricks left over from the construction of the house. I ran an electric cable in there, plugged a bare bulb into it, and started laying the bricks, with no mortar, to form a tiny room. I left a small opening to enter and exit from. I built a fort and it felt safe.

I soon turned it into a darkroom. We were half a mile away from a shopping center and I would bike there all the time. I began to buy the basic supplies for a wet darkroom. I got red and amber lightbulbs, trays, and some chemicals, then began to print contact sheets with a pane of window glass. Every time I washed out the trays, I had to crawl out of that hole, jump down and use the basement sink. I shot black and white Plus-X Pan film, 127-format, and made only contact prints at first. My dad bought me a small enlarger. It was amazing to watch prints materialize in front of my eyes.

17

CALCULATORS, POLAROIDS AND CASSETTE DECKS

I WENT THROUGH A SHORT PHASE OF BUILDING PLASTIC model airplanes in the basement of that house. I built World War II fighters and hung the completed ones from the basement ceiling rafters with wire. There were no electronic devices in this room, aside from the washer and dryer. The technology in that house was the same as it was in Yonkers. Consumer technology had changed little in the ten years I had been alive. And if it had, it made little difference in my life. I knew that some technology was old-fashioned and that some was more modern, "space age" as we said. Still, my life was not substantially different than my brother's life had been, ten years earlier.

My inner life expanded. Books helped me to cope with being alone. Comic books helped me to escape from reality. I was a fan of both DC and Marvel, but I was an omnivore and read widely. I read Doc Savage paperbacks after my father mentioned them to me. I read EC Comics when I could find them. It was in Princeton where I got my Dick Tracy anthology, which featured the hero's futuristic crime-fighting tools. These once seemed as modern as James Bond's high technology a few decades later. I got a thick hardcover anthology of the collected comic strips of "Buck Rogers In the 25th Century." These strips took me away every day on imaginary journeys into space, as the original Star Trek television shows had done a few years earlier. Science fiction did not seem so improbable. If you could think of it, scientists would eventually invent it. That's how I saw it at the age of ten.

The private school I went to in Princeton was nicer than P.S. 28 had been, but classroom technology remained the same for me until the day I walked into a small room with an IBM computer terminal. This was something new. The Basic and Fortran languages were alien to me. I might have given them more of a chance if I had not been 10 years old. I knew this computer was the future in some way, but I had no idea what that meant in practical terms. It didn't seem revolutionary to me at the time because I saw nobody doing anything special with it, aside from creating what people would later call "ASCII art." This computer did not change my life, but it was like spotting something fly past the car window. Your head would turn, your eyes would register it, and you might subconsciously file it away in your mind. You knew that that you might see it again one day.

What did blow me out of the water was the Bowmar electronic calculator my dad brought home one day in 1971. He came home from work, pulled it out of a small vinyl pouch, turned it on and handed it to me. This was a tangible object from the future. It was tiny, silent and made instantaneous calculations, which it displayed in ruby-colored LED characters. Without knowing a thing about chips or printed circuit boards, I knew this was something different. Nothing in my life up to this point had combined so little mass and so much power.

For a kid who did not love math, the Bowmar was a glorious display of what technology could do for me. It was my dad's, but he let me play with it and it made an impression that does not diminish, even after 50 years. I would turn it on and watch it spring to life without warming up, then feed it huge calculations, which it performed instantaneously. It could perform only the four basic arithmetic functions, but that was enough.

My father had been using a slide rule for decades, well into the 1960s. He had tried to impress me with the elegance and power of this device, but to no avail. I never bothered to learn how to use a slide rule. This calculator was a revelation, though, and my father knew it, too. He quietly retired his slide rule to a storage cabinet and never used it again after 1971.

My father's brother, George, who had been shot at and shelled in the Battle of the Bulge, had a looser attitude on life than my father did. He was more into gadgets for their own sake. Aside from the Buicks that he owned, fully loaded with every option, he also had an electric juicer, the first I had ever seen, which co-existed with their mother's steel bread box from the Great Depression and rope-operated dumbwaiter in their ancestral tenement apartment.

George and my father had a mutual friend, named Sam, who owned the first Polaroid camera I ever saw. This was another big deal. It was tangible, in-your-face technology. The Polaroid was dramatic enough to take over any party or event. Sam would snap a few photos, pull them from the camera, let them sit for a few minutes, then rip the paper off, exposing that glorious chemical smell to the room. I would catch a glimpse of the photos while he rubbed something on them, some fixative thing that looked like a tube of lipstick, then people would pass them around and marvel at the snapshots. This was instantaneous, magical, and real.

The photos that my mother was taking were now color. She abandoned black and white. My father had been shooting color slides for much longer, and those ancient Kodachrome slides still glow, decades later, a window onto the past.

The slide shows ended for us by 1970. Color negative film took a while to become mainstream, but it took over. Roland was about to discover the ancient, hand-held, hand-wound eight-millimeter movie camera that was stashed in his house, but he and I lived 90 miles apart at the time and my experience with that camera would not come until my family moved back to Yonkers.

While I was living in Princeton, Roland and I each got our own small Panasonic cassette deck, the kind with piano key control buttons and a small wired microphone. Cassettes were another revolution. Any time a gadget shrank a great deal, it made a big impression on us. There was nothing so tangible as miniaturization. I would record a rambling, pre-adolescent monologue onto one side of a tape and send it to him in an envelope through the U.S. Mail. Then, I would wait for him to listen to it, record himself on the other side, and send it back to me. We did this a few times, but it was more fun and gratifying to make a phone call.

At this early age, each of us was preoccupied with capturing and manipulating sound. The cassette was a natural extension of the reel-to-reel. You could see the little reels in the enclosure. It was easy to understand. We didn't know about demagnetizing the tape heads or cleaning the transport system, but we instinctively knew how to fix a mangled tape. You had to get a pencil and carefully wind it back. You might also have to splice it together with cellophane tape.

Another marker of technological progress was the evolution of Scotch tape over the course of the 1960s. It went from a thick, yellow sticky tape to a clear, hazy version with less tack to it. You could still buy the older version, but the new one was considered more modern. By 1970, ballpoint pens had become common. When my brother was in school, ten years earlier, Schaeffer fountain pens were the norm. The big improvement for him had been plastic disposable ink cartridges. Instead of refilling the pen by squeezing the bladder reservoir, you popped out the old ink cartridge and put a new one in. Ballpoint pens were much less of a hassle. They would sometimes dry up, at which point you could hold the ball tip under a lighter for a few seconds to get it running again. There were also stories – accompanied by showing a horribly stained shirt or sport jacket pocket – of ballpoints exploding on planes. By the 1960s, Pentel felt pens, Magic Markers and ballpoints were surging. Mechanical pencils were fading.

One item I treasured came from my uncle's ad agency. It was an electric pencil sharpener. You stuck in the pencil and it would automatically start to grind and sharpen the point. Every once in a while, you would have to empty the wood shavings from the little hopper. That machine felt like a big improvement over the small, manual crank-operated sharpener that was screwed to the side of my bookshelf. Electrification meant progress.

In Princeton, television signals still came over the air, music still came over the radio or a vinyl record, and the telephone was still hard-wired into the network. We got a new area code.

Direct dialing of what were called "long-distance numbers" was new to me. If you were local, you did not need to dial an area code. This was true for much of the country into the twenty-first century. In the 1970s, many cars still used leaded gas. Nobody carried bottles of water around. Even in the early 1970s, there was far less plastic than you would find in later years.

From the little knurled plastic top of the "3-In-1" oil can to the occasional plastic part in a toy, the material stood out more for its rarity than for its ubiquity. Toys were made out of metal until the 1960s. By the 1970s, that had begun to change. Paper cups became plastic cups. Paper bags became plastic garbage can liners. We used the brown paper bags that you got at the supermarket. You spilled out your cup before you dumped it in the trash. Otherwise, it would leak all over the place. The first plastic trash bags we used were for leaves. Before that, we would burn our leaf piles. I don't remember soda in plastic bottles before the 1970s. It was all cans or glass bottles until then.

When I rode my bike to the stationary store to buy comic books, I was alone with my thoughts. The store smelled like newsprint and ink. Toll booths were manned by humans, except for the occasional one where you would throw coins into the metal basket and drive on. Cash registers in stores were mechanical or electro-mechanical. Nothing was networked. There was no GPS. People used maps and road atlases. This was the early 1970s. We knew that better technology was out there. NASA was proof of this. Humans had gone to the moon and sent back a live television feed. Even if the new technology had not trickled down to us yet, we knew it was on the way.

18

YONKERS AND THE ARCADE

In 1973, after two lonely, existential years in Princeton, we moved back to Yonkers and into a house on the other side of the dead end where I had grown up. I was thrilled to be back in the old neighborhood. It was 1973 and I was thirteen years old. The new rock music sucked. It felt as though every musician had died or burnt out. The 1960s ended as I was coming of age. That decade had been a magical time to be a child, but I became a young adult in the 1970s. It coincided with a change in my consciousness, partly due to adolescence and partly to drugs, but also due to technology.

The telephones in the new house still carried the network over copper wires. They now used a "modular jack," also called an "RJ connector," instead of the big four-prong plug connected to the phone line. I squinted into one of those modular plugs and marveled at how tiny the wires were inside the clear little cube with the plastic clip. We had a phone in my parents' bedroom and one in the kitchen. My father also ran a line to the big room downstairs.

The garage door was electric, as were the kitchen stove and the oven. The coils on the stove would heat up and slowly turn red. The new house in Yonkers had a bigger dining nook, but the same small TV and radio. This nook had a picture window onto the backyard. We had a dining room that we never used. The attic had a tremendous fan mounted on the floor. When you turned it on, louvers would open in the ceiling below it from the air suction. In theory, if you opened the windows on the lower floors, air would get pulled up and the house would cool down. It never worked that way, but it didn't matter. This house had central air conditioning, which was new to us.

As in Princeton, we still got our television signal over the air. All of our neighbors did, too. Cable TV was not available. In this house, we got our first color television and it went into the den on a wheeled stand next to my father's desk. Once the color barrier had been breached, there was no going back. They replaced the dining nook TV with a small color set. The den had two easy chairs, one for my dad and one for my mom, and a wall full of books. His stereo remained the same and he added a cassette deck to the rig at some point. Everything else in this house remained at the same level of technology as before. In the basement, my father and I put up a thin wall, using studs and wood paneling, and I caulked up the cracks to create a proper darkroom. It had no running water, but at least I could stand up in it, unlike the crawlspace in Princeton. I stopped shooting film after a while.

On some significant occasion in my early teens, probably a birthday, my dad solemnly took me to the local Radio Shack and we purchased a tuner/receiver, a turntable capable of stacking 7 records at a time, and two large speakers. All components were the Radio Shack house brand – "Realistic." I had studied their catalog carefully and was familiar with the whole model range. I knew I would have to settle for something at the low end of the lineup, but I lobbied, unsuccessfully, for the slightly better models on all three components. I got the boxes home, hooked the system up in my bedroom, and immediately started to blast records. My father must have been having second thoughts downstairs. My own system!

My record collection began in Princeton, where I would go to the university bookstore and browse. I had little money and little sense of what I wanted, but I did buy some LPs while I lived there. Most of my money went to comic books and paperbacks. Back in Yonkers, I raided the metal milk crates that held my brother's records, his ancient, worn-out monaural Stones and Beatles records, his discarded Jimi Hendrix, Flying Burrito Brothers and Byrds albums, and these became the basis for my own record collection. He did not seem to miss them, so I got a big leg up on my collection with these two milk crates of LPs. I entered another school and encountered pot, all within the same few weeks.

I was a bomb waiting to explode, a future degenerate and drug addict about to happen. Every night, I would go to bed with the old Koss headphones on my head after carefully placing a stack of seven LPs on the turntable. I would start the first record, get under the covers with a fifteen-foot extension cord running from the amplifier to my headphones, and disappear into a world of music and fantasy. If I were still awake after the seventh album ended, I would get up in the dark, tiptoe to the turntable, raise the dust cover, flip the whole stack, then get back into bed and continue my psychic explorations. I always fell asleep with the headphones on.

After a year of this, that I discovered record cleaning brushes. I took care of my LPs. They were, along with the printed covers, objects that my friends and I handled and studied for hours, from the photos and design to the liner notes. A cool record might even have an inner sleeve worth studying. Millions of kids across the country were doing the same thing, imbibing the same cover art and messages at the same time.

My bedroom was plastered with posters I bought in Greenwich Village at Bleecker Bob's record store. My discovery of pot coincided with the introduction, in 1973, of the Bic lighter, a cheap and handy device that delivered thousands of lights. This was handy when you smoked marijuana by the single hit. Roland and I collected our spent Bic lighters in an old fishbowl in his basement, whose walls were also covered in posters from Bleecker Bob's.

Most technology stayed the same for me while I was in high school. Our television had no remote control. I used a manual typewriter for my homework assignments and my watch was a self-winding Seiko that I wore infrequently. Handouts in school were still mimeographed. When some friends and I began to publish a satire magazine, I learned how to type on a mimeo stencil and feed it into the machine. I had been introduced to photocopiers in the late 1960s, when I accompanied my dad to his office for a day, but I did not get to play with one until the 1970s. Photocopier technology felt like magic. We called them all "Xerox machines," whether they were made by Xerox or not.

Carbon paper was how I made copies of school papers. "Wite-Out" and scratch erasers were how I corrected them. I handed the stapled papers to my teachers in person. I took notes with a pen in notebooks and studied by memorizing things, using home-made flash cards. This was how most students worked until the 1990s. It's what the technology allowed. When we talked, it was on the home telephone. The New York City subway cost thirty-five cents when I was in high school. It was cheaper when I was younger. You dropped a brass token into the slot on a turnstile for one fare. Some of the subway cars in the system dated to the 1930s, with windows that opened, no air conditioning, wicker seats and small, oscillating fans.

I learned to drive in high school. My father had let me sit in his lap when I was much younger and steer the Rocket 88. We would go on dirt roads in the Catskills and he let me take control of the wheel. Driving felt intuitive. When I hit my teens, he bought a car with a manual transmission so I could learn how to shift. He wanted me to be familiar with that technology. It may have been a Pontiac.

We would go out and drive on the steep hills of Yonkers, where I learned how to start a car with a manual transmission on a steep hill. He carefully explained – with hand gestures – how a clutch worked. My brother would give me extra tips on how to "toe-and-heel" with a clutch – using the brake pedal, clutch and gas pedal all at the same time. Driving changed my relationship with machines. I had to learn to adjust the machine to my body every time I took control over it, adjusting the seat and mirrors before I put the car in gear.

During one of those endless high school summers, my parents made me take a touch-typing course at the local high school. I hated the idea and did not want to do it, but they insisted. I have been grateful for it ever since and I am touch-typing these very words. I sat there every day for two weeks in front of a manual typewriter. A grizzled old guy, sweating in the heat, solemnly chanted the letters and we taught our fingers the dance, etching into muscle memory the QWERTY keyboard. It's amazing how my brain and hands were able to learn how to type without looking. The methods now may or may not differ, but the object is still the same. Current wisdom says that cursive handwriting is dead. Could better voice recognition technology make touch-typing obsolete at some point?

One day in 1973, just when Roland and I were discovering weed, rock and roll, and girls, we walked into the pizza place on Central Avenue and Tuckahoe Road, in Yonkers. We were probably looking for trouble and didn't know it. We got our slices and stood over a glowing object – a table-based arcade game that was unlike anything we had ever seen before – Pong. We sat across from each other and played a round. This otherworldly thing had fallen into our lives from another dimension. We put in all the money we had and played for as long as we could, mesmerized by the glowing electrons hitting the cathode ray tube. The game itself was secondary to the sensory experience. It was simple to play, yet difficult to master. The action kept getting faster and it turned us into zombies, as we concentrated completely on the screen. Between the sounds and the glow and the physical playing, it was a perfect experience. We kept going back to play. A revolution was under way and we were part of it.

We were regulars at Adventurers Inn, further north on Central Avenue. Adventurers had long rows of pinball games, as well as shooting games that spat out a tiny target sheet afterwards – an actual piece of paper with puncture holes in. There were tiny bowling games, where the ball would curve upwards on the alley, and electro-mechanical arcade games with controls that resembled periscopes, tank turrets or racing car cockpits. The arcade was a hurricane of noise - people shouting, guns shooting, sirens blaring, bells ringing, and screaming conversations, all at the same time. "One more...PLEASE????" "Can we eat now?" "CHECK THIS OUT!"

Outside were bicycles, most laid on their sides and not locked up - Stingrays, newsboy bikes and racers. Cars were parked in the paved lot, where young kids hung out and smoked cigarettes. Inside, people walked back and forth between the arcade and the massive food section – a brilliant pairing of enterprises. Adventurers Inn later changed hands and became an outpost of Nathan's Famous. Very little changed inside, though. It had a block-long dining area with large Formica picnic tables. A stainless-steel tray rail ran the length of it. You would enter at the start and walk in one direction, passing different food along the way and ending up at the cash registers.

This was cafeteria style fast food at its best. Behind the glass were steam counters full of hot dogs, fried chicken, and cottage fries that were greasy and delicious, among other foods. You paid with cash and then went directly to the huge condiments bar. When you were done eating, you walked through a small square opening in the wall, where the netherworld universe of the arcade assaulted you. Old cadaverous guys and desperate-looking younger guys walked around in ugly uniform shirts. They had no future and they knew it. They wore heavy aprons, each one with three bulging pockets at waist level filled with quarters. They made the change for you with dead eyes, then you went back to whatever game was the next on your radar.

Roland and I developed hand-to-eye coordination and learned how to use body English. We each had certain games which we were better at and we avoided the others. I had to choose wisely. I had little money and my parents thought the games were a complete waste of time. We set out from home on our bicycles and, since we were not allowed to go to Central Avenue, we pedaled hard to get there fast. We had fun for half an hour, then rode those three-gear bikes up steep hills for a few miles to get home. Doing it on the sly was half the fun. We did that for years and continued doing it when we were older and able to drive there by car. The parking lot at Adventurers was a hangout for the local pseudo gangs, but we went there for the games.

Soon after we first played "Pong" in the pizza place, Nathans began to replace many of their electro-mechanical games with video games. There were not many at first. Within a year or two, I remember playing "Death Race 2000," in which you ran people over with your car and left little crosses or tombstones in their place. That made a big impression on me. They made only 500 of those games, one of the only video games to NOT contain a CPU, or central processing unit, and then they recalled them all. They were too unsavory for the time. When you look at the carnage and bloodshed in later generations of video games, you have to wonder what changed so much in society.

The strip of shopping centers along Central Avenue also contained a General Cinemas movie theater. There were other theaters in the suburbs near us – in downtown Yonkers, Bronxville, Mount Vernon and the Bronx, but we usually went to the General Cinema. Movie theaters were special. The picture and sound were way beyond anything available at home and the communal experience was also fun. Going to a movie was an event. I also liked the popcorn and candy.

1973 and 1974 saw me getting high on weed for the first time. I loved it. Once the inhibitions were gone, I was doing pills, cough syrup, and anything else I could lay my hands on. All of this recreational drug usage made me appreciate good stereo sound more than ever, as well as interesting visuals. Visuals were difficult to find, though, unless you went to a rock concert. By that time, the golden age of performance visuals was already over and the age of the VJ had not yet arrived. We were stuck in the suburbs with limited resources. We bought colored light bulbs, black lights and black light-friendly posters. We may have even had a strobe, but I remember only a flashing light and not its source. Either way, we were looking to enhance the acid trips with technology. TV was the only entertainment available. Videotape was not yet a home technology.

Roland found an 8-millimeter movie camera that had been stashed in his house since the late 1950s. You would load the film, wind it up with a thumb key, focus the fixed lens and then shoot, planning each shot with care. We made a few 3-minute films with no edits, simply by pausing between each shot. They were more weed-induced explorations than linear narratives. He had to send the film out to be developed and, as always, it was a thrill to get the reel back from the processor and watch it for the first time.

19

BIRTH OF A GARAGE BAND

I WAS PLAYING GUITAR MORE AND MORE. IT BEGAN WITH A hand-me-down, cheap, nylon-string classical guitar from my brother. After a few years, I bought a plywood Japanese acoustic. My parents saw that I had no intention of quitting music and they were dismayed. "Musicians are all bums, Paul. Why do you want to be a musician?" This came from my mother, who loved music and respected musicians, but did not want her son to become one. She may have seen enough hopheads while working at Decca records during the Bebop era. Or maybe she was just trying to scare me.

She did not succeed. I wanted an electric guitar, but I had to work for it. I worked and saved for it the whole summer. Roland bought a Univox "Hi-Flier" guitar, a copy of the Mosrite "Ventures" model, along with a solid-state Gibson amplifier. Our jamming began in the 1960s on plastic guitars with fake strings made of nylon fishing line. Now, we could play bad folk music and crude, two-chord rock. I was smashing out chords on the Japanese acoustic and Roland was learning solos on his Univox. This was the stuff that garage bands are made of.

It was 1977 when I got my first electric guitar. Electric guitars reflect technology as much as anything else, and it felt good to move into the modern age. They are also the gateway drug to electronic effects and recording equipment, as well as amplification. This guitar was a Gibson Melody Maker reissue, meaning it copied a design that had originally appeared fifteen years earlier. It was a bottom-of-the line student guitar, but it was well-made in Kalamazoo, Michigan at the same Gibson factory that turned out the legendary Les Pauls, Flying Vs, ES-335s and other Gibson guitars.

I purchased a Peavey solid state amplifier that was loud enough to get me in trouble with the neighborhood, even though it had only one 12" speaker. It had a wired footswitch for overdrive and distortion. You would pre-set just how dirty you wanted the effect, then hit the button with your foot and briefly become Jimi Hendrix in your head. When I turned this amp over and peered into the back, there was a small sticker on the bottom of the reverb unit that read "Assembled by Beautiful Women in Meridien, Mississippi." I was duly impressed.

With these two guitars and two amps, planted in Roland's basement, we could now pump out loud, rough versions of Stones and Hendrix songs. We also played other staples of the mid-1970s, including scary electric versions of the folk songs we had been performing in the public parks and train stations of Yonkers on our acoustic guitars.

Our first gig was at Gerde's Folk City, in Greenwich Village, in 1975. We played a 9-minute instrumental original that was long and rambling, but so earnest that we got sympathetic applause. Once we had the electric guitars, we could sculpt with much bigger chisels. We could also tinker. I added another pickup to the Melody Maker, a Gibson Firebird pickup that I bought at Bronen's Music, in White Plains. Roland eventually bought a Les Paul Deluxe for $400 at We Buy Guitars, on 48th Street in Manhattan. We played and played and played. We smoked weed and kept on playing while the colored light bulbs burned.

Our setup was crude. We sang at the top of our lungs over the two guitars. Roland was more diligent, learning solos and licks from records. I practiced, but not nearly as hard. I was awful for a long time, but we had fun. Simply carrying an electric guitar around in its case gave me reason to swagger. We felt like rock stars. Plugging a cable into the guitar, then into the amp and tuning up was a spiritual experience each and every time. It still is. Amplified sound from one's fingertips began for me in the 1970s.

We began to tape record our jams. When we were done playing, we would drive to a parking lot on Central Avenue in Yonkers and listen to the tapes while we smoked weed. We would drive all the way to the emptiest corner of the parking lot, turn the car around so that we could keep an eye out for cops, then light up and hit the "Play" button on his tape deck.

The technology of my high school days was no different from what my brother or even my father had experienced in school. My high school, unlike the school in Princeton, had no computer. Personal computers did not exist then except for hobbyists. A mainframe cost a lot of money. My high school had a great art room and it was a refuge for me. I never learned how to throw a clay pot on the wheel, but we had a lovely, empathetic woman who taught me how to screen print. The technology used amber gelatin film. You cut it with an X-Acto knife and melted this into the silk with acetone, in the process destroying your cuticles, and then you would peel off the backing paper and have a finished screen to print. Printing was done with ink and a rubber squeegee. I loved making prints.

I designed posters for the high school theater and glee club with rulers, pencils, Letraset and Prestype. These two products were press-on type, used by art directors and layout people for creating text and copy. You would lay the soft plastic sheet on the paper, line it up, then rub the letter you wanted to adhere. It was satisfying to create a body of type, but you had to be slow and meticulous or it looked sloppy. "Desktop publishing" was not yet a concept, let alone a phrase.

If I could not find a typeface that worked in the Letraset sheets that I had, I would find one in a type book or on a record album, then use tracing paper to copy the individual letters. If the album didn't have the right letters, I would create them from bits of the existing letters. It was a good education in the elements of typography. I would eyeball spacing and kerning, draw it on the tracing paper, then cut the stencil from that. The only technology involved was the fluorescent desk lamp that I used. I was building type by hand, with limited means, and working in a technological vacuum.

The fluorescent drafting lamp clamped onto the side of my desk. It looked like every swing-arm, single tube desk lamp in every movie about Strategic Air Command or on the desk of every accountant in a film noir. It was 1940s technology at its best. The biggest change in technology between me and my father was in quantity. There was more stuff available to me. His family had been poor. We were middle class, so I had a radio and a phonograph in my room, rather than one for the whole house. There was no phone in my high school bedroom, but nobody I knew had a phone in their bedroom at that time.

In 1977, Roland and I discovered electronic guitar effects, another harbinger of the space age for us. He was the first to buy one of these pedals – a Fresh Fuzz. This was an overdrive unit, happily known as a "fuzz box" to most musicians. You plugged a guitar into one side of it with a quarter-inch cable, ran the output cable into an amplifier, and then got nasty. It would turn a guitar, in adjustable increments, into a dirty-sounding beast with endless sustain. Surely, this was progress. Roland then bought an original Dunlop "Cry-Baby" wah-wah pedal, which Jimi Hendrix had immortalized. This had a potentiometer, essentially a knob, that you controlled with your foot to make the guitar express itself vocally while you played it, hence the "wah wah." Another victory for the future.

I earned enough money at shitty jobs to buy a Small Stone Phase Shifter. This let me dial in gigantic sweeps, swooshes, swirls and sonic whorls. I could add texture and depth and adjust the speed of the effect. It was not the most useful device, but it was fun. I then bought an Electro-Harmonix Micro-Synthesizer, which had 10 sliding switches that let me create unlimited audio shades and textures. I could shape the sound in strange ways, climbing one more rung up the ladder to outer space.

We worked our asses off and saved money until we were able to buy a Peavey RX-200 PA system, a monaural unit with 5 inputs and two big speakers. We also bought two good microphones. It was a huge commitment and it was also the biggest and most expensive thing either of us had ever purchased. With all of this gear, we were ready to conquer the world. We put the PA system Roland's basement. We would set up the guitar amps and, even before we started to play, would revel in setting up the gear - the plugging-in of cables, the placement of microphones, the glow of the power lights, along with the buzz, crackle and hum of the amplifiers. It was a joy to feel the power of the electricity. Then, we would begin to play and our mortal limitations were audible to us and to everyone else, but at least we could strive towards something, we could dream and, occasionally, we could transcend.

Our neighbors may not have seen it that way. Mr. and Mrs. Glynn were Roland's next-door neighbors, and their house was less than 10 yards from Roland's basement window. Mr. Glynn had always been nice to us when we were kids. He had been in the U.S. Navy and seen combat in World War II. He had a kapok-filled life jacket in his garage from his navy days. He let us play with it when we were little kids. It was dirty and gross, but we appreciated it. He was a good guy.

We had grown up with him as a neighbor, but our roles had changed and we were now the pot-smoking, trouble-making teenagers next door. Even worse, we were making a lot of noise. Roland's downstairs phone would ring – the old Western Electric 302 – and we would stop playing and look at each other. The magic mood would quickly end, even though the colored light bulbs continued to burn. Roland would answer the phone, only to have a distraught Mrs. Glynn shouting "Mr. Glynn is trying to sleep!" We would then turn down the volume and try to keep it down for the rest of the session. It did not feel like the spirit of rock and roll, but we tried to obey the neighbors. The spirit of rock and roll, though, would manifest in us doing exactly the same thing the next day and the day after that. That rig, the guitars and the P.A. system, may have been the most modern things that we owned.

20

PERSONAL SOUND

MUSIC WAS EVERYTHING TO US. WE WERE LUCKY TO HAVE the little mono cassette decks, but we knew we could do better. We had little money, though, and good tape decks were expensive, so we subsisted on what we had. By senior year in high school, Roland was able to buy a portable TEAC stereo cassette field recorder. It was about fifteen inches long by nine inches high and four inches thick. It felt like it was right out of "2001 – A Space Odyssey," with its big, round, illuminated audio meters on the front panel. We could carry it any place we wanted to go and record anything we wanted to record with the big stereo condenser mic he also got.

I bought a mono Panasonic deck, an improvement over my childhood tape deck. It had a tiny, rectangular analog meter which functioned, but which I never trusted. There was a quarter-inch output, for headphones, and the sound quality was good. The little cassette door was dampened, so it would open slowly without a cheap clacking sound. This felt like progress. We taught ourselves the rituals of cleaning and de-magnetizing tape heads. We learned about recording levels, and we became master labelers of tapes. It seemed like half of cassette tape culture in the 1970s and 1980s was about labeling the tapes.

Cassette tapes were the first way to customize your audio experience on a song-by-song basis. You would sit up late at night and painstakingly make mix tapes. First, you made sure that the record had been dusted and was clean. If you had a direct drive turntable, you would adjust the pitch control and make sure the platter was spinning at the correct speed. You did this by looking at the little strobe light on the side of the platter and making tiny turns with a tiny knob. Then, you would cue up the tape.

You did this in one of two ways. You could hit play and watch the tape leader go around half a turn, then pause it. Or you could do it manually, by turning the spool with your pinky or a pencil. Once the cassette was cued up, you would hit the "Pause" button, then press the "Play" and "Record" buttons at the same time. You set the proper sound levels by gently putting the needle down in the middle of a loud-looking section. How did you know what a loud section looked like? You looked for rougher grooves in the vinyl. Or, you could just plop the tonearm down a few times while checking the meters. If the levels were in a comfortable range, you were ready to go. You might compulsively clean the record again at the last minute, then put the needle in the groove, wait a second, and quickly un-pause the tape deck.

It was always a good idea to listen to the whole recording in real time, just in case the record had a skip and you needed to redo it. If you were recording albums, you would let the whole side run and turn up the volume near the end to make sure you didn't cut off the last notes. Then, you paused the tape deck and let the tone arm pick itself up and return to its rest. You would flip the record, clean it, put the tone arm down again, un-pause the deck and record side two. Most albums fit neatly on one side of a ninety-minute tape. Forty-five minutes was the yardstick. If you had time left over, you could fill up the rest of the tape side with an extra song or two. Double albums were usually done consecutively, so you would often have a good amount of tape left over at the end of the second side. These extra spaces were never left empty and it was fun to fill them in a creative way.

While the record sides were being recorded, you would label the tape, writing the information on the tape labels for the box and the stickers for the tape, itself. Depending on how much weed you had smoked and how compulsive you were, you could get very focused on making a clean, legible label, sometimes with multiple color pens and extra information. We all had our favorite labeling pens. I liked fine tip felt markers from Pilot. Maybe it was the influence of the weed, but I liked the way they felt as I wrote. I liked the clean lines and the color palette.

Making the tapes was a meditation. When you listened to the tapes later, over and over again, they would take on a certain physical patina. The boxes would get scratched and the tapes would degrade in miniscule ways, but they were rarely redone if you did them right the first time. Some music might become less interesting or you might re-record something onto a better grade of cassette. In this case, you would record something different over the old music on the old tape. I never threw a tape in the garbage. I would keep a few rolls of typing correction tape on hand and re-surface the tape labels so I could re-use them.

Compilation tapes were put together on a song-by-song basis and I pulled these off the radio. It's hard to record a song from the radio. DJs talk over the beginnings and endings of songs, unless it's a jazz station. Still, radio was the only place to get some music at the time. Roland carried a good assortment of cassettes, about twenty at a time, in a low cardboard box under the seat of his car. Once he installed the first tape deck in his Opel Manta Rallye, we were in business. Our main form of leisure, after playing guitars, was driving around Yonkers, listening to music and smoking weed.

There was no Sony Walkman in the mid-to-late 1970s. Vinyl was king. Cassettes were what you had in cars. Radio was the great form of mobile technology, but nobody was walking around with headphones on their head. You might see somebody with a transistor radio and a tiny mono earphone, but not often where we lived. We listened to music at home or in the car. My parents' car did not have a cassette deck. We did not get an AM/FM car radio until 1979. When I first began to drive the family car, I had to listen to the AM stations, where it was hard to find good music in New York in the late 1970s.

Those car radios were analog. The tuning knob moved a small gear and wire mechanism. This physically moved a red marker that told you where you were on the dial. You saved a station with one of the eight push buttons. The radio dial was lit with a tiny incandescent bulb and the single speaker was in the middle of the dashboard, facing upwards. Our 1973 Oldsmobile had an antenna embedded in the glass of the windshield, rather than a metal rod extending from the side of the car. I thought this wire in the glass was super cool. The rear window had no defrost wires embedded in it, so my father bought a strange little fan that he installed in the back of the car. He screwed it into the panel beneath the rear window. It hooked into the electrical system and he wired up a switch to turn it on or off. It made a lot of noise while running and never cleared the fog off the rear window.

The 1970s were my teen years. When it came to technology, my biggest windows into the future were the cassette decks and the arcade videogames. There were no personal electronics yet. It was not much of a concept, although I'm sure there were people in Japan and California dreaming about devices they might build. Small televisions and radios were being sold as personal, portable devices, but not something you would carry on your person. At the time, "portable" meant that it had a carrying handle.

Any decent spy movie featured miniature devices that packed a lot of punch. You could see answering machines, based on tiny spinning tapes, built into a wall. Agents would have handheld communication devices and exotic weapons. You could read Dick Tracy and see him talking to people on his wristwatch radio. I felt as though I were stuck in the provinces. I did not know that people had been walking around with Sony Portapak tape recorders for half a decade already, making fuzzy video recordings. My access to technology news was limited to "Popular Mechanics" magazine. At the time, transistor radios were as portable as it got for me.

21

FLIGHT

As for transportation technology, all I had ever traveled on by my mid-teens were bicycles, automobiles and the occasional ferry boat. I never flew until 1976, when I was sixteen years old and my high school glee club did an exchange program with the country of Poland. We flew there that summer and that's where I was on July 4th, when everybody in the U.S. was celebrating the U.S. Bicentennial. We flew on a Boeing 707, the first pervasive commercial jet.

We sped down the runway at JFK and became airborne. It was like losing my virginity. When I took that flight to Poland, most of my classmates were veterans of flying, so I had to act casual, but I was excited. We watched a movie on the flight, with a projector beaming the film onto the screen at the front of the cabin.

Poland, 30 years after World War II and still part of the Communist Bloc, felt like a technological backwater compared to America. I came upon a vending machine in an underground walkway that dispensed some fruity water into a glass, which you would then replace, to be used again. We toured the country and performed for people in concert halls and in public places. We saw Auschwitz and the Majdanek extermination camps, which stay with me to this day. We drove on a modern bus and ate greasy food. The Poles were proud of having rebuilt from the rubble after World War II. We flew back on a 707, as well.

Roland and his family had been flying to Florida and Italy for years. He would describe how cool it was to roll down the runway at a few hundred miles per hour and I would have to imagine what that felt like. My grandparents flew to Florida and Greece often. I loved to go pick them up or drop them off at the airport. They would return from Florida with sacks of oranges and grapefruits, which we had to snatch from the amazing luggage carousels. Everything about flying was futuristic and magical to me, from the TWA terminal at JFK to the big jets sitting outside the terminal windows.

My father traveled often for business and he would bring me souvenirs from the planes. I loved the overnight kits that he picked up on his flights to Asia. People dressed up for a plane flight and treated it like a special occasion. Flying felt more civilized and elegant then. You could accompany your family and friends to the gate. You did not have to prostrate yourself before the security screening process. Air travel was something to be excited about more than something to be endured. My father, the veteran business traveler, probably did not see it that way, but I would bet that his first flights, on DC-3s in the 1940s, were exciting for him.

It was all about progress. I gave a "current events" report in the fourth grade on the new plane that Boeing had just announced, the 747. The specifications were monumental, from the number of seats, to the length and wingspan, to the double-decker sky lounge. My teacher, Mrs. Warnock, remarked how she had been a stewardess on DC-3s and had to wear high heels. She remembered how steep the aisle was when the plane was on the ground. The DC-3 had a tail wheel instead of a nose wheel and she said that walking up the cabin in heels was like climbing stairs. The Douglas DC-3 had revolutionized air travel. The Boeing 707 did the same thing for the jet age.

The 747 was in my future, but time moved slowly when I was young. Four or five years then was a third of my life. I remember calculating, when I was eight, how long it would take before I could get a learner's permit to drive. It was six or seven years in the future and that felt almost immeasurable. The Concorde was flying by the time I was 10 years old and my friends and I all waited for the day when America's own SST, or "supersonic transport" would take to the air. The idea of one day enjoying space travel was not the impossible comic book fantasy of my father's generation. We knew it would happen. It was a matter of "when" and not "if."

22

TOUCH TYPING AND FACTORY JOBS

It was not until my senior year in high school that I got an electric typewriter. Up until then, I used a manual model, starting with my mom's black enamel Royal, which she had owned since the 1930s. By the 1960s, it was an antique, but she soldiered on with it, one sheet of paper at a time, one ink ribbon at a time. If you wanted something to look professional or if it were important, you typed it. You had to be careful. If you made a mistake, you had to erase it. This often wore a small hole in the fibers of the paper and made permanent smudges.

You learned how to apply "Wite-Out" with finesse and skill, not heavily, letting it dry before you typed over it. You might also use "Ko-Rec-Type" correction film. These were thin sheets of plastic coated on one side with a white emulsion. You back-spaced to the mistake, stuck the sheet on the paper, then typed over it and the Ko-Rec-Type would usually white out your error. If you were typing a rough draft, you could just "X" the mistake out. If it were a final copy, though, it had to look good.

This was the state of technology for me as I finished high school and prepared to attend college. I had gotten accepted into Harvard College and, while I could not pass it up, I was ambivalent. Our garage band had become important to me. It was 1978 and the primitive, violent music we played was now being played in small clubs all over downtown New York City. We were young, single and completely immersed in our music. There was nothing we enjoyed as much as playing all day in the basement. Such a time and opportunity would not come again, but I didn't know this when I was young. I had some forgettable corporate intern job that summer, but it didn't last.

My previous jobs had been menial. At the age of 15 or 16, I worked at Waldbaums, the local supermarket, as a price changer. I had to wear a bright blue polyester jacket and change the prices on items when they were raised or went on sale. You could not just slap a new sticker on the items. You had to remove the old price first. My supervisor drilled this into my head. If it were ink, you would rub it off with some Windex and a small rag. If it were a sticker, you had to scrape it off. After you removed the old price, you stamped the new price on with a purple ink stamper or you used a plastic price gun to put the new sticker on.

It was mind-numbing idiot work of the first magnitude. The worst items were cans of cat food or tomato paste. I would sit there for hours, usually after closing time, robotically removing items from the shelves, cleaning the old price off, putting the new price on, and then re-shelving the items. It was an early wake-up call for me that a great machine existed whose purpose was to suck me in as a tiny cog, a worker drone. Interaction with other humans might have staved off this dark realization, but I was alone in a grocery aisle with a lot of time on my hands and a vivid imagination. I would later discover the same ethos in office jobs. It felt as though society wanted to kill my spirit. Later, of course, UPC bar codes would render price changing jobs extinct, but it was my first taste of real labor and I hated it.

That job lasted a month. After that, Roland and I worked at a refrigerated furniture factory in Mount Vernon, a shop where they made desks with refrigerators and secret bars built into them. They once had us unload a freight car of refrigerators from a railroad siding nearby. We weighed less than each refrigerator. The owner looked at us one day and said, "you guys are like two shits in a bowl." We were there as a favor to Roland's grandmother, so we could not complain or talk back.

We moved on to a small factory in Nyack that made perfume and cosmetic products. It had gigantic mixing vats and reminded me of a Batman comic book. They had us string wire from the ceiling for the factory's public address system. They had us move boxes. In both jobs, we were deemed to be useless labor, bad human stock. We excelled at nothing and would often pilfer something useful, like a box of solid brass screws. We would collect our first checks and never return. At all of these early jobs and even a few later in life, I punched a timecard in an electro-mechanical clock to keep track of my hours.

I had no plan or goal for what I wanted to do with my life. My subconscious plan was probably to get through school and then to keep playing with the band, but I was clueless. Fellow students of mine from school charted their lives out and were working to make their dreams happen. I was living in the moment, in a perpetual haze of garage band music and weed. Going to college was a cop out. The more ballsy thing would have been to leave home and find a way to make it on my own in the city. I must have thought I could go away, get a degree, then return and play music again - an endless teenage summer. Nobody pushed me hard on what I wanted to do, so I followed in the grand tradition of getting a degree in English, something I already knew well. I would live a dissolute life, then get a job in some drone-like profession and play music. At least I could live on my own at school.

I spent the summer of 1978 recording cassettes, playing with the band, and taking long bike rides up to the Kensico Dam in Westchester. By this time, I owned a Raleigh "Gran Prix" 10-speed bike, which I had rebuilt a few times. I was familiar with every part on that bicycle and had adjusted it to be absolutely silent no matter how hard I pedaled.

I would bring the mono Panasonic tape recorder to school for my audio needs. The stereo would stay at home, as would the bicycle. I was supposed to be going away to work hard. The only serious technology I would take to school, aside from the tape deck, was my typewriter. I packed clothes, a few books, some posters and a hot pot, that mainstay of dorm cooking technology. I had a single room waiting for me and was ready to leave home.

23

COLLEGE

THE DAY CAME, IN SEPTEMBER OF 1978, WHEN MY parents and I drove east for 200 miles with a roof rack full of crap. We rolled through the rear gate of Harvard Yard, opposite the Science Center. After the obligatory unpacking, conversations, exhortations and neurotic exchanges, I found myself alone in my room. For the first time in my life, I was under no supervision.

I put in a request to have my phone connected. This would become a rite of passage each year. You put in the request and New England Bell would hook up the phone a few days later. In a few weeks, you would get a bill in the mail, another milepost on the road to adulthood. You would then call home and give the phone number to your parents. Some schools had a public phone or a small phone booth in the hallway that the whole floor shared.

Personal computers, the internet, Wi-Fi and cellphones did not exist. You could go to a payphone and call home collect, but you were cut off from anybody back home until you got the phone hooked up and called people with your new number. This created a big disconnect and break with your past. Now, you keep your phone and can call or text your friends and parents. Going away to school then felt like you were really going away. It was not strange for me to write the occasional letter and send it. When you didn't want to communicate by phone, letters were useful.

I got my first credit card with the Harvard Coop, the Cooperative Society. It was the college bookstore, as well as a large department store. This card was my plastic gateway to a record collection. It was also my first real taste of debt, another marker on the road to financial slavery. I got a bank account with Baybank, who gave me my first checkbook in life. My only previous experience with banking had been a sturdy paper bank book, which I used as a child to amass less than 50 dollars.

The Baybank card was blue and green and I could use it at the ATM they had just installed in front of their Harvard Square branch. This was the first year Baybank offered an ATM, another sign of the modern age. Now, you could get money without going into the bank, signing a check or a withdrawal slip, and handing it to the teller. You no longer needed a human to get money. This was remarkable. That ATM would not accept deposits, so I still had to see a teller when I got a check in the mail or a paycheck. I used the Baybank card whenever I had money in the bank. This machine dispensed five-dollar bills, which was handy for somebody who was chronically broke. ATMs would figure into some sordid episodes later in my life and this is where it all began.

One way to blow off steam was to head to Tommy's Lunch, where they had great sandwiches and video games. I would buy a Raspberry Lime Rickey, change a few bucks for some quarters, and settle in front of an arcade game. Tomorrow's creative, intellectual and financial leaders spent whole days and uncounted quarters playing these early video games – Missile Command, Zaxxon, Millipede and others. We were at our physical peak in terms of hand-to-eye coordination. That and the cocaine ensured long sessions of wasted time, moving on to the next level in each game.

The first job I got at school was at the University Copy Services. My boss at the time was an overweight pervert with long, dirty blonde hair. On my first night at work, he asked me, in a heavy Boston accent "Have you ever been to an orgy?" I was creeped out. I stayed there for another week, though. I would show up at 9:00 p.m., pick up a folder with the copy jobs, then wheel a hand truck with boxes of copy paper to a remote basement somewhere across campus, where I ran the copy jobs.

I fed the originals and blank paper stock into huge photocopiers under bright fluorescent lights, smoking weed the whole time and choking on copy toner fumes. The machines were bigger than refrigerators and they had daunting control panels, along with modern plastic surfaces. They felt futuristic at the time, but now seem quaint, and they made a lot of noise. These giant copiers were machines from the industrial part of the data revolution.

The worst copy jobs involved staples. You had to separate the originals with a staple puller, make the copies, then staple the originals back together again when the job was done. In the early hours of the morning, I would wheel the hand truck back, boxes full of finished copy jobs, then I would pass out in bed and skip class. That job was bleak and existential, but the next one was even worse.

I became a student security guard and wore a blue shirt with an official patch sewn onto the left sleeve. I walked a prearranged route and, at regular spots on my rounds, I would stop, open a small metal "Detex" box, turn my key in it and then move on. Somebody checked the boxes on a regular basis to ensure that the security guards were making the rounds and not sleeping on the job. That job was too bleak for me, so I quit and found work in the Harvard dining halls, where I could talk with people while I worked. I have had to learn this lesson more than once in life. No matter how much I think I'm a lone wolf, I need people and human interaction. Another lesson I learned was to schedule classes only in the afternoon.

School devolved into a vortex of drugs and alcohol. I showed up and functioned, but at the minimum level of effort required to get by. I showed up for exams, I ate in the dining hall, I consumed and sold drugs, and I listened to a lot of music. No matter what was going on, a record was spinning nearby on a turntable. The dining hall work felt degrading. I stood at a waist-high window and took dirty trays from my fellow students. I requested a position away from the tray return window. They put me at the pot-washing sink, another job I loathed.

I quit and took a job at the local grocery store, where I manned the cash register, stocked groceries, scooped ice cream and baked cookies. It was at this store where I saw an odd piece of technology for 1980. It was a video jukebox. You dropped coins into the slot and a short, pre-MTV music video played for three minutes. Nancy Sinatra's "These Boots are Made for Walking" was the most popular video on this twentieth century artifact.

It was here, one December night in 1980, as I was closing the store, that I heard about John Lennon's murder. I was perched on a small brick ledge, pulling the security gate down, when I heard one student on Mount Auburn Street calling to another "Hey, did you hear? John Lennon was shot." He probably heard it on the radio. I heard it shouted on the street. There were no personal alerts or notifications. When major news occurred, you turned on the TV or the radio to find out what was happening. CNN and the 24-hour news cycle were born in 1980, but most people did not have cable. If a big story broke, broadcasters cleared the schedule and ran special coverage.

24

COLLEGE RADIO

ALL THE LOCAL RADIO STATIONS PRE-EMPTED THEIR programming and played nothing but Lennon and the Beatles for the next day or two. By that point, I had joined WHRB-FM, Harvard's radio station, and that was where I devoted most of my energy for the final two years of my college career. I had to get an FCC license and pass a proficiency test in the studio at WHRB.

Everybody at the station took on a handle to use in the station logs. I chose "AXIS" in honor of Jimi Hendrix. I broadcast as many radio shows as possible. I loved being in the studio in the basement of old Memorial Hall. The front door was down some steep steps at the thin end of the building, and it felt like you were entering an ocean liner through the prow. It was a glorious cave. Inside were an office, a newsroom with an AP ticker, fed by rolls of sweet-smelling newsprint, a few studios, a couple of record libraries, and an ad-hoc lounge. It was the first place in college where I felt as though I belonged.

The main broadcast studio had two turntables, two very old, but solid Ampex reel-to-reel tape machines, and a rack holding a large patch bay, the broadcast signal meters, and amplifiers. The turntables sat on a mixing panel with big round fader knobs. When I first started, the turntables were ancient, but they were eventually replaced with modern Technics units.

I began my broadcasting career with a show from 3:00 a.m. to 6:00 a.m., once a week, for the Department of Great Black Music, formerly known as the "Jazz Department," but renamed in the late 1970s. They let me play anything, although some of my musical choices did not derive strictly from the jazz tradition. I might start a show with Eric Dolphy, then segue into a few minutes of Lenny Bruce, then Henry Cow, into Jimi Hendrix and then Bird. Free-form music pumping out over the airwaves – terrestrial radio. The transmitter was located on a building not far away.

I talked my way into a 12:00 to 3:00 a.m. slot with the rock department, doing reggae music with a heavy emphasis on dub, a fairly new genre back then. I had been turned on to dub by a Jamaican guy I worked with in the dining hall. We both washed pots at the same large sink. He lent me a copy of "Pick A Dub," by the Second Street Dreads, and "King Tubby Meets Rockers Uptown." This music, forged in the studio on mixing boards, was heavy on echo, spacey effects and deep bass. It hit me hard.

I would travel to every record store I could find, searching for dub albums. Before online music and before Amazon.com, much of the joy in being a music fan was the hunt for record albums in stores. My friend, Paul, and I began to do a show together. We called it "Midnight Dread" and built up a small, dedicated following of like-minded reggae fans and space cadets from a radius of 30 miles around. We quickly figured out how to route our broadcast microphone through the patch bay, where we would create a live tape feedback loop through one of the Ampex decks. We then run it back into the board, allowing us to lay a heavy, adjustable echo on our voices. We had a lot of fun echoing our breaks. It took little to make us happy.

The technology we used was ancient. WHRB had no money, despite being Harvard University's radio station. Some of our gear dated from the 1960s. The people in the Tech Department were treated like gods because they kept us on the air and kept our broadcast signal within acceptable technical standards. They repaired the old equipment and maintained the transmitter. Our facilities may have been crude, but we had a commercial broadcasting band and enough power to reach many people.

WHRB tolerated a high level of eccentricity and degenerate behavior. Paul and I would smoke weed all night in the studios, filling up the whole facility with the smell of strong sinsemilla. In the morning, one member from the classical department would invariably come in and blow a gasket. We didn't care. We treated this show as a mission, a crusade. We reveled in making interesting, auditory mind-fuck segues, in playing long stretches of strange and interesting music, and in taking as many requests as we could. The station had no cassette deck in the broadcast studio. The engineers didn't consider cassette tape audio quality acceptable for broadcasting. At the time, much of our best music was on cassette, copied from rare records or priceless tapes that collectors had given us.

At the beginning of my sophomore year, I bought a stereo system from a small, second-hand stereo store on Mount Auburn Street. I'm not sure a store like this could exist today, but it did then, and it thrived on the lively market of college students who bought and sold old equipment. I bought a Nikko power amp with an impressive number of knobs and controls on the front panel, as well as a respectable power rating. Its previous owner, I was told, was a member of the Boston Symphony Orchestra with the euphonious name of Fenwick Smith.

I bought an old Garrard turntable with a white, baked-enamel surface and a belt-drive. Back then, the big arguments were over the relative merits of belt vs direct drive turntables. I bought a new Aiwa cassette deck and two well-used, but powerful speakers. I installed this rig in my tiny dorm room, ran a cable to the living room and hooked up some cheap speakers I had pulled from the garbage. My roommate, Peter, and I now had a serious sound-system and we never turned it off. The little lamps in those components were always glowing.

From that point onward, whenever I moved in life, after the last box had been loaded into the room and the door finally closed, the first thing I did was to hook up the stereo system. I had to have sound before I could unpack. My power amp had outputs for three sets of speakers, and I would run wires to different rooms. Then, with the push of an electro-mechanical switch, I could send a signal to remote locations. The joy I got from this little trick never grew old.

25

THE WALKMAN

My other equipment purchase, in the fall of 1980, was the original Sony Walkman. I don't remember how I heard about the Walkman, but I knew I had to have one, even at the hefty sum of $200. I worked overtime to buy one and I remember inspecting it for the first time, the way it felt in my hand. The surface was brushed blue metal and it seemed shocking that something so small could play a cassette. This was another big leap forward in miniaturization. It could not record, but it would play full-sized cassette tapes. It came with a small set of headphones and, thoughtfully, two output jacks. Roland and I took long bike rides that summer up to White Plains with the Walkman strapped to the rear rack on my Raleigh Grand Prix. I wore one set of headphones and he rode no more than 10 feet behind me the whole time, wearing another set of headphones with a long cable.

It was all we could muster at the time, but it was the first taste we had of personal electronics and portable media immersion. We would play everything from Jimi Hendrix to Brian Eno's "Music for Films." The original Walkman had an orange button on the top. This activated a built-in mic whose sole purpose was to allow two people using the same Walkman to talk with each other while wearing headphones. You pressed the orange button to mute the music and talk. Your friend would hear your voice in the headphones, then you would let the button go and the music would resume. We never used this feature and it disappeared from all future portable music players, as far as I know. Sony was still experimenting with what people wanted.

The Walkman allowed me to dub cassette tapes to my big Aiwa cassette deck. This was a great advance in capability. I could now make tapes and duplicate them for others. The Walkman also let me play cassettes on WHRB without having to drag my big deck down to the station, which I had been doing with a backpack. I would patch the Walkman into the board with my carefully guarded stereo-mini to dual RCA plug cable. If the engineers had known about my setup, they would have disapproved, but it made no difference to the broadcasts, which sounded good enough to me. I made airchecks by leaving a blank cassette in the deck back in my dorm room and having my girlfriend hit "record" when I went on the air.

Within a year, Aiwa came out with their own tiny deck, and this one could also record tapes. Roland and I had been dragging his heavy TEAC deck to concerts, along with the big stereo mic, in order to make live recordings – "bootlegs" – of shows. We recorded some amazing shows this way, but it was impractical and slightly dangerous to smuggle such a big deck into a concert hall or club. The Aiwa made surreptitious recording practical.

Whenever we got together and went to shows, he brought a small microphone and we would stand there, casually holding the tiny mic in front of us. We could now take the music home. We carried spare batteries and tapes, and we made sure the cassettes got flipped every 45 minutes. We tried not to scream or talk near the mic. We knew we could never release the music and were dimly aware of copyright issues. We did it out of love, as fans.

We called our venture "Axis Productions," again in homage to Jimi Hendrix. I had a tiny rubber stamp with the planet Saturn on it and we would stamp the cassette label with it and then write, in small letters circling the planet, "An Axis Production." We would label the tape "Original Master" and make a copy for each of us. It was a logical step from recording live shows to playing them on the radio. Late at night, I would play the shows we had taped without announcing any details.

At some point, I met a young guy from Boston who collected Jimi Hendrix material. He had a large number of obscure shows, bootlegs, board tapes and other stuff we had never heard of before. He brought some of his material by the station for me to play, but Roland and I wanted to own it, so we made a deal and paid him hundreds of dollars for the right to copy his tapes.

Years later, real collectors that we ran into disparaged this guy and his way of doing things. It was considered wrong to make people pay for this stuff, but we were desperate for the material. Roland sat down to the hard task of mass duplication. After five days of non-stop dubbing, we had our own collections. In the pre-digital era, everything had to be copied in real time. If a tape was 90 minutes long, that's how long it took to copy.

At the radio station, getting a smooth segue with the Walkman was not easy. If you pressed the "Play" button, you got a loud pop when the head hit the tape. Even if you paused it on "Play" first, the segue was rough. The ideal method was to cue the tape at the right spot, pause it with the board fader down, then release the "Pause" button with one hand and simultaneously turn the fader up with the other hand.

It became body memory, as did all the little moves that you made with turntables when cuing up a cut on vinyl. There was a physical component to making a broadcast from that studio or any studio, for that matter. You were part of the machinery. Live DJs and sound people, as well as recording engineers, still serve that function. You are the main chip on the logic board. Machinery and technology are designed to interact with the human body, mind and senses.

26

THE LOOP

My academic technology at Harvard was rooted in the early part of the twentieth century, even the nineteenth century. Pens, spiral notebooks, and typewriters were the main tools I used. We had Widener Library, with its magnificent floors of underground stacks – miles of books on shelves, punctuated with tiny desks at hermitic distances from each other. The card catalogs were all paper based. There were micro-fiche machines, a technology I had encountered in the Yonkers Public Library, but which I did not use much.

Microfiches and microfilm impressed me - anything miniature impressed me - but they were a logical technology. You could understand that, through optical wizardry, an image could be made tiny and then read with a magnifying machine. Cool, but understandable. It was not a technology that made you say "Whoa! How did they do THAT?"

There was little other technology at school that was new to me. Some friends were taking "computer courses," but I had no interest in that field, nor did I see how it might work in my life. Computers were something you might use to crunch data for science, but I saw no place for them in creative work, let alone everyday life. Neither did most other people, unless they were computer scientists or visionaries. All of that would change in less than five years. This was just before the first great evolutionary thunderclap of personal computing.

At home in Yonkers, on breaks from school and during the summer, the band ground on. We picked up a drummer and suddenly became a loud trio. We were happily writing songs and playing, mainly in our basements and at the occasional party. We continued to record our work and listen to it in Roland's car in the huge parking lots on Central Avenue.

It was satisfying to hear and critique our own work, which we had written, played and recorded. We never said it out loud, but we felt as though we were on the front edge of music and technology, at least at the no-money, D.I.Y. level of music. We knew our limitations. We knew there was technology out there that was beyond our reach. We were happy to have what we had and that was good enough. It was certainly good enough for rock and punk music.

During one of these long summer days at the start of the 1980s, Roland and I were happily sitting around his basement, surrounded by all of the sound effects, guitar cables and amplifiers that we owned. We ran a signal from one effect to another and then another - a tangle of cables that terminated in one channel on the PA system. We ran this signal out of another channel and back into the first device on the chain. We were just hooking up devices to each other. Decades later, neither of us could remember why.

Nothing happened, so we sat back and talked about music. We had been chatting for about 30 minutes when the reverie was broken by a distinctive electronic beep, followed by a grinding noise, a glissando, then a series of spitting noises in a cascade running up and down. We were electrified and looked at each other with wide eyes. It sounded like a hidden maestro at the levers of an electronic orchestra.

The feedback loop had come to life on its own and was transmitting otherworldly patterns and noises. We pulled out the tape deck and recorded 30 minutes of it. We felt as though we had discovered a new planet. We got out a small notebook and documented exactly how all the cables ran and where the knob settings were. We called it, predictably, "The Loop" and spent the next few weeks editing and naming snippets of what we had created.

Even though we knew about the world of electronic music, it was not until we made our own that we became obsessed with it. We had been driving Roland's Opel down to Inwood, at the northern tip of Manhattan, to buy weed from a taciturn guy who ran a record store on Dyckman Street. We usually bought an LP every time we went to the store. Until then, we had been filling in our collections of rock and jazz, but now we wanted any electronic and experimental music we could get. Beaver and Krause, Henry Kaiser, Mother Mallard's Portable Masterpiece Company – these are some of the artists we found and explored in that store's bins. It was hard to find anything like this on the radio, so we had to seek it out in small record shops.

We made many trips to that record store and we created more electronic music. We tried to revive the loop a few times, but it was difficult to trigger and control. It seemed to have a mind of its own, so we let it go. We did other things, though. We used effects, feedback, double-tracking, tape decks and any other form of analog trickery that we could harness. Roland bought a Revox reel-to-reel tape recorder and we did some proper double tracking. This was how we spent the summers.

27

THE PARTY CONTINUES

DURING MY SECOND YEAR IN COLLEGE, I WAS IMMERSED one night in a double-tracking experiment with two tape recorders. I had been on a double date with a woman who lived on my floor, along with a good friend and his girlfriend. We all took acid, listened to music, and drifted away to our own worlds at the end of the night. I was recording directly into one tape deck and then bouncing tracks while recording into the other one. I was three bounces into it when the woman from my floor knocked. I took off my headphones and we talked. She stayed the night and became my girlfriend. I got no more multi-tracking done for a long time.

Roland and I first became aware of Sun Ra and his Arkestra in 1979. They played straight jazz in their own way, but Ra was also one of the great innovators of "space music," electronic music, experimental, and free jazz. His musicians dressed up in costumes, capes and gold outfits and put on a show that was a cross between Coltrane and a Séance. We were blown away by the Arkestra and caught as many shows as we could, recording as many as possible. These years shaped our lives and sensibilities. Anything was possible, sonically and literally, as long as the technology was there. When Ra and the Arkestra began to chant "You're in the Space Age Now," we knew exactly what they were talking about. We expected more to come. We didn't pay as much attention to the next line: "There's no place that you can run."

So continued the great party that was my college education. Minimal schoolwork, much time at the radio station, playing music in all my spare time, following bands around the Northeast, and doing copious amounts of different mood-altering drugs, from pot to speed to coke to acid to psychedelics of indeterminate origin. I used to regret not taking full advantage of the premiere education that was offered to me, but I have made peace with that regret. I explored other frontiers. I would pay the price later on, rather than collect the reward for a hard-won education, but I was not thinking about tomorrow. I was already living in tomorrow.

My experience with advanced technology up to this point was in creative pursuits – experimental music and radio broadcasting. In schoolwork and employment, I was still living in the technology of the 1960s and, in some cases, the technology of the 1800s. One of my first summer jobs in college was an internship with the Mobil Oil Corporation. The pay was bad, but it was an opportunity to get real-world experience with an old-fashioned, paternalistic corporation, something that would lead to a safe and secure career. This was how my parents saw it.

I saw it as a total waste of time, a living death. I had no illusions about what it meant to work for a big company, what it meant to work for anybody but myself. I have always considered it a complete waste of time, something to do as a last resort, only when things got desperate, which they often did.

I worked in two different departments at Mobil. The first was in the Public Affairs department as a fact checker, where I made sure that Mobil's weekly Op-Ed piece in the Times did not stray far from plausible deniability. This job was easy enough for a young English major. I showed up every day and they gave me busywork. I did not appreciate the value of busywork at the time. What I did appreciate was the horror of getting on the train every day at Tuckahoe Station, in the suburbs, and heading to the city to surrender my time for a paycheck. This is what my dad had been doing for decades. I had nothing but respect for how he lived his life, but it was not what I wanted.

The allure of New York in the late 1970s and early 1980s was enough to distract me from the misery of actual employment. I put on my shirt, tie and jacket, tucked my chin into my chest, and gutted it out, spending vast amounts of energy trying to look busy. I had an IBM Selectric typewriter, which eased the pain a bit.

The Selectric was the penultimate evolutionary stage in typewriters. It was big and heavy. When you turned it on, by flicking a little rocker switch under the front edge, the machine would quietly hum. If you hit the "Return" key, the carriage would jump and the platen would rotate with speed, power and authority. Typing on it was fast and smooth. It had an autocorrect button, in case you made a mistake. This would backspace at the touch of a key and lift off the carbon impression you had just typed. You could type it again and create perfect copy.

Even more impressive was the ball, or "typing element," which replaced individual typebars and made the letter impressions on the page. It was fun to watch the ball dance across the page. If you had optional replacement elements, you could change the font of what you were typing. The Selectric made every other typewriter in my life seem old and dull.

Mobil had a modern, 1970s phone system, but I never used it. The secretaries' phones could handle multiple lines. They were like miniature switchboards, where you could transfer calls, put people on hold, or create conference calls. These phones were analog. They performed all of these functions with switches and wiring, rather than printed circuit boards and chips. Most of the desks, including mine, had single-line phones.

My job in Public Affairs did not last long. I was transferred one day. I don't remember why, but it could have had something to do with my pothead-level of performance. My boss sent me to the department of corporate records. This was an important office where most of the company's corporate papers, as well as deeds and property documents, were housed in a small room. It was a bleak place, presided over by a woman who had been there for decades and worked her way up from the position of file clerk.

She showed no mercy for buffoons like me, people she could sniff out as time wasters and slackers, who cared little about whether they made a mistake or not. The records were stored according to an index card system that lived in steel filing cabinets. She tolerated no fools and brooked no carelessness. She was absolutely faithful to the file system and to filing perfection. Any drone from the business world of the previous century would have been able to pick up the work with no trouble, so embedded was it in nineteenth-century filing technology.

I hated it. I would get a stack of documents to file each day and it had to be done in a precise, specific way. As I put papers away, I would see fascinating glimpses of history – articles of incorporation for small holding companies, deeds, letters of ownership – the company's history, the industry's history told through its original documents. I did not have time to enjoy them, though, and there was no way to copy them for later perusal. I was a corporate drone and the job became grist for one of the punk songs we wrote at the time in Roland's basement. I needed the money to buy new equipment, but the work drove me insane.

The ID card I used to gain entry into the building every day was plastic. There was no magnetic stripe on it, nor was there any information embedded in it – that idea was still radically futuristic. It contained a name and photo and you held it up for the guard to see. This was considered high security at the time. The only reason they had this system was because the Mobil Building had been bombed a few years earlier. Most other buildings at the time, as well as airports, were open to the public.

28

RADIO RESEARCH

In the summer of 1981, I got a job that was just as drone-like in many ways, but more interesting and laid-back. There was a radio station in New York at the time with the call letters "WYNY," on the FM dial. It later became an awful country/western station. When I was there, it served up pop music. This job fell under the auspices of the "Radio Research" department at WYNY, but it was far less scientific than the name would suggest.

I sat with my fellow researchers in a small, windowless room deep inside the RCA Building, three of us at a time. It was a rotating crew of young people, but we never bonded with each other. The room could hold no more than four people. There was a Formica counter around three walls. It held four workstations, each with a filthy office chair and a telephone. The phones were standard issue, but each had a hold button and access to the other phone lines. The access was with pushbuttons - the old-style, clear round pushbuttons at the bottom of the keypad. A simple cassette machine hooked up to each phone with a mini jack.

We made cold calls. At Radio Research, we called people and tried to convince them to give us five minutes of their time, once a week, for ten weeks or as long as we could retain them. We would establish a relationship and then call to have them rate songs that we played in thirty-second snippets from the tape recorders. The songs changed slightly every week, based on the ratings and position in the pop charts.

It was difficult to find people willing to rate songs for us. We offered them nothing. We hooked enough people to stay busy, though. We preyed on the lonely, the bored, and the elderly. I don't remember the script that we read them, but I do remember the monotony of dialing numbers for hours at a time and the numbing horror of listening to the same 30-second clips of bad pop songs all day. "Hello, Liz? This is Paul from Radio Research" was my standard salutation.

The draw for all of us was access to the building. The supervisor, who watched us like a den mother, was a kind of super-fan. He knew all the celebrities and loved working in the RCA building. He had been at this shitty job for years and still lived in New Jersey with his mother. There was cachet and glamor in getting onto the employee elevators at 30 Rock, then called the "RCA Building." That was the only perk. We were part-time workers with no benefits, low pay and no future. I left in two months, as I had no access to the studios or any of the things that interested me. This was just another dead-end, low-tech job on the fringes of media.

Later that summer, I bought a short-scale electric bass from Bronen's music, in White Plains. It was a Fender Musicmaster and was great for a guitarist who wanted to play the bass, but not spend time mastering it. I would learn later in life just how brilliant Leo Fender had been in his invention of the electric bass. He revolutionized music with the Precision and Jazz basses. We plugged the bass into guitar amps, even though we knew it might blow out the speakers, until we bought a used tube bass head, by Earth Amplification. It was point-to-point wired - each connection was made with a wire soldered to another part of the board. There were no printed circuit boards in that amplifier. Even though it was built with vacuum tubes, that amp worked for decades with no maintenance.

29

NEW YORK CITY - 1982

AFTER FOUR YEARS OF INTERMITTENT STUDY, LOTS OF radio broadcasting, and many psychedelic excursions, school ended. I packed up my books, along with the stereo system and the Ohaus triple-beam scale, and returned to Yonkers, New York, to my parents' house, to the band, and to the sudden reality of a future I had not thought to plan for. It was June of 1982 and I was 21 years old.

The woman I had met in college and already lived with for two years stayed in Boston with her parents. I moved back in with my parents and immediately began to play with the band, which reconstituted itself again. It was not an ideal situation, but it could have been worse. My girlfriend and I went to Europe as a way to stave off reality a little longer. We had no money, but were determined to go, so we bought Eurail passes and flew to London.

The London transit system was on strike, so we could not take the tube, but we took buses. We hitchhiked to a Stones show at Wembley and caught a ride with a car full of English punks. We took a ferry across the channel and rode slow trains all the way to Greece, sleeping on the cold deck of the boat from Brindisi to Piraeus.

This trip was low-tech to begin with, and it was made even lower tech by our relative poverty. We telephoned home once and it was a logistical hassle. We ran out of money at our furthest terminus – Iraklion, in Crete. We placed a desperate call home for money, which our parents wired to us via Western Union. We sat for a week in the main square, drinking bathtub ouzo and running up the tab at a cheap hotel, waiting for the cash to arrive. We had spent the travelers checks weeks before. We had no credit cards.

When I was a kid, credit cards were less common. They were not rare, though. My father had a fake alligator skin wallet in the glove compartment of the car that held at least five different oil company credit cards. The companies offered them to promote brand loyalty. My mom had cards for Wanamaker's Department Store, Lord and Taylor, Gimbel's and a few others. Credit cards looked the way they look now, but without a magnetic strip or a chip. Their main function was to maintain and transfer the account information - the embossed account number and expiration date. These cards did not have CVC security codes, either.

You paid for something and the salesclerk, waitress or gas station attendant would make an impression of your card by putting it in one of two machines. The more common one was a rectangular device, about a foot long, with a moving handle that enclosed a hard rubber roller. They would place the charge slip, in triplicate with carbon paper between the two copies, on the metal bed, then grab the handle and pull the apparatus over the card. This would push the raised plastic characters into the paper and carbon paper and make a perfect facsimile of your card info. It was like making a gravestone rubbing.

The other machine operated on a similar principle, but it had a single push lever that moved the roller. At some businesses, the prices were put onto the slip by little machines that looked like adding machines, but most businesses used a ballpoint pen to fill in the purchase information. They would separate the paper copies – the customer receipt, the sales receipt, and the copy for the credit card company - and their bookkeeper would treat it as another piece of paper in the accounting chain.

The earliest kind of credit card fraud was thieves stealing garbage from businesses, then taking the carbon paper, along with all of your card information. Carbon paper was replaced with carbonless copy paper, a technology that impressed me, but bigger changes were ahead.

The only modern technology we carried on that European trip was my Walkman. I built a portable speaker from a small wooden cigar box and a speaker pulled from an old transistor radio. After adding a plug from Radio Shack, it was a serviceable extension speaker. I carried a few cassettes on that trip and had some epiphanic moments, listening to tapes of Bob Marley and the Wailers, along with Nino Rota, Blondie and Sun Ra. The speaker made it all the way to the island of Rhodes, where it finally disintegrated. I buried it near a beach.

We made it back to Athens and took a 747 home. School was over and I had no plans. I began to play with the band again and, prodded by my parents, looked for work. I took the expedient of working for a temp agency, which was a way to avoid committing to anything serious. The current wisdom was that you could move from a temp job to a permanent position. It was a viable career path but, in my case, temping was a way to tread water.

While I was with this temp agency, I encountered my first word processing machine, the Wang word processor. I liked the Wang. It did not take long to learn how to use it and it felt like the future. There was something comforting about the big terminal, the heavy-duty keyboard, and the glowing green phosphor letters on the screen. By today's standards, it was primitive. You could navigate within a document only with cursor keys. There was no graphical user interface, and the screen could display only ASCII text. Still, there was something magical about it.

Most of my temp jobs consisted of answering phones, using an IBM Selectric typewriter, filing papers by hand, and making copies on big photocopiers. The Wang let me create perfect copy. It let me keep a document on a big disk and return to it later. At another level, it let me disappear into a glowing screen, to interact with a cathode ray tube in a way that I never could with a television. It was more like a video game. It was not networked, it was not that smart, but it was satisfying and fun to use.

I did what I had to do at these temp jobs, then I did what I could do for myself. That meant creating song lists and charts for the band on the Wang. I would save these onto five-inch floppy disks, each of which held one hundred kilobytes of information. I held on to these disks for years, long after I was able to recover the information on them.

In addition to talking by phone, via landline, my girlfriend and I would correspond with written letters. It was more intimate. We had done this for the past few summers. It was always a thrill to get the day's mail and find a letter from her, a giddy little joy that demanded to be read immediately. For the important stuff, though, she would call on the telephone, and that's how she delivered the ultimatum that would catapult me from the comfort of my parents' house. "You have to find a place to live in the city this month." It was basically "get us a place to live or it's over."

I came out of my stupor, worked a bit harder, and started walking around downtown Manhattan with a copy of the Village Voice, real estate ads circled, every day after work. I found a sub-ground studio for rent on far West 23rd Street. She moved down and I discovered a hard lesson in life: two people living together should have at least one door that they can close at any time. It was small and sweet, but we needed another room. Still, it got me out of the house, got her out of Boston, and gave us a home base in Manhattan.

It was a six-month sublet and it cost $500 a month, which seemed like a lot of money, but we made it work. Moving was easy then. I had few possessions. The most important things I owned were my records, milk crate after milk crate full of records, hauled around from room to room, from sublet to rental. When I moved to New York City, I took a small technological step backward, as my apartments all had cheap refrigerators that would choke up with frost and need to be defrosted, unlike the "frost-free" unit I grew up with. As a young man with no patience, I defrosted with a hammer and screwdriver.

I got a job around the corner and put my Ivy League education to work as a night porter at the Empire Diner, on 10th Avenue. I was below the busboys in rank. My job was to clean the floors, take out the garbage, and to do any low-level chore that needed doing. One of the waitresses, a dissolute young woman in her early 30's who was always high on coke or downers, saw me and said "You're the porter? You're the first guy we ever had on that job who speaks English."

It was a simple life, and I would roll out of bed at 9:00 p.m., walk around the corner, head for the basement and work until 6:00 a.m. Everything there was transacted by hand, and the most advanced technology was the jukebox. I felt as though I had arrived at some new level in my life, unsure exactly what that was, as I ate my free meal of scrambled eggs at 3:00 a.m. while the waitress danced around the kitchen drunk, singing "Stupid Cupid" along with Connie Francis on the jukebox.

That job did not last long and I went back to temp work. My girlfriend got a good job with a public relations agency and I followed her there. My life fell into darkness for the next seven years. I had struggled in the past, overdosing on alcohol at the age of 13, using too much pot in high school, then harder drugs in college. I functioned at a low level, but I managed to graduate. By 1982, I had gotten clean and was just smoking weed and drinking. Then I moved into that little studio in West Chelsea in 1982 and got back into hard drugs. The 1980s would become my lost decade.

30

THE REVOLUTION

By 1982, when I moved to New York City, many people had begun to buy and use home computers. Personal computers appeared earlier. In 1977, Radio Shack's TRS-80, the PET and the Apple II arrived. I never encountered one in college and nobody I knew had, either. By the early 1980s, the Commodore 64 and various Atari machines were widespread. People began to use them in their homes.

The public relations agency began to buy personal computers for the administrative assistants. They used the IBM 5150 model. I knew it was more complex than the Wang machine, but I had no interest in learning how to use it. I saw it as a glorified word processor. Trying to use it felt like I was staring at the algebra textbook again with my father, the engineer, staring at me from across the table. These machines had no graphic user interface. They used DOS – "Disk Operating System" to operate the machine. MS-DOS, the operating system that IBM paid Bill Gates and Microsoft to develop, looked like a wall of white text on a blue screen.

That first IBM PC was effective, but not intuitive to me. It felt complex and arcane and there was no way I thought I could master it. I was too distracted by the drugs and depravity in my life to focus on anything. Keyboard commands were beyond my ken at the time.

Apple came out with the first graphical user interface in 1984, on the original Macintosh. The creation myth around this operating system is that Steve Jobs and a small team of Apple engineers visited Xerox's PARC research lab, saw what the future was, and then went back to Apple and built their own version. That's essentially true. Apple arranged a deal and compensated Xerox with stock options so that they could inspect the Alto computer.

The future they glimpsed had an operating system that was visual. Instead of white text on a dark background, a user saw the representation of a desktop. You interacted with the computer visually, rather than with lines of text and keyboard commands. This was a fundamental paradigm shift. Instead of punching in a few characters from the keyboard to delete a document, you clicked on it and dragged it to the trash.

The Apple team got the computer mouse from Xerox. It helped you to navigate the desktop visually. The Xerox lab had 100 Alto machines networked together and this was revolutionary, as well, along with the ability to print a document from the computer to a printer exactly the way it appeared on the screen. The Apple team had to go back to their own lab and write every line of code from scratch to make this possible, but they pulled off this feat and the Macintosh Operating System was born.

I wanted only to play music and get high. One piece of technology that did thrill me in the 1980s was my Casio watch. It felt like an advanced device. Anything with a digital readout felt modern. Early digital alarm clocks had mechanical numbers that would flap over as the time changed. The Casio watch was based on liquid crystal display technology and the numbers silently, fluidly changed. I studied each section of each number and admired how they could create any combination by turning the different crystals on or off. You would set the time and date with a couple of buttons. There was even a tiny LED light that you activated by pressing a button. I would lie in the dark and turn on the light, just to bathe in the tiny glow.

My father, with some ceremony, had given me a self-winding Seiko in the early 1970s. It had a stainless-steel band that always caught the hairs on my wrist and pinched my skin. It was heavy and reliable. The coolest part was how it wound itself with an internal weight and a spring, as my father explained it to me. It was all mechanical and a beautiful piece of engineering, but the Casio was lighter, waterproof, and nearly indestructible. The Casio bands always broke before the watch did. I would go to Canal Street to get cheap replacement bands. The watch was scratched and scarred, but it never failed. It was inexpensive and it had four functions – time, date, alarm, and stopwatch.

31

PUBLIC RELATIONS

There was plenty of video production work at the public relations firm. They had an in-house producer and we put together compilation reels of news clips that the agency had generated. We edited these on videotape, which was not common outside of broadcasting in the early 1980s. Public relations firms needed in-house video, though.

PR is a bullshit game moving in two directions. You have to convince a legitimate media outlet that you have something newsworthy that can credibly fill up their surplus of airtime. You save them work by offering a story in the guise of news. They play along and allow your spokesperson to slip a product mention into the story. This gives your client the implied endorsement of a television network, radio station or print outlet.

The second part of the equation is show your client how much value they got. You have to sell your results back to the client. If it's in print, you make a scrapbook – clippings of all the "hits" where their message appeared. If it's radio, you make an audio compilation tape. For video, you make a video reel.

They are all trophies. Three-quarter inch tapes, along with VHS and Betamax tapes, came to us from the TV stations. If we could not get a free tape, we bought one from one of the broadcast monitoring services that record everything at all times. The tapes came in and I had to string the segments together. It was the perfect job for a pothead.

The hardest part of this job was trying to appear cheerful. I worked under a guy who had been a hack news producer for a major network. He still had connections, so the head of our PR company kissed his ass. In return, he would get important clients placed on network shows. It was an ugly business all around, with one hand washing another, but all hands remaining filthy.

This producer was never friendly to me and my disdain for him was clear, but my girlfriend was a rising star in the agency, so they put up with me. The producer would book studio time and assemble tapes with clean cuts. Desktop video did not yet exist. He would finish off these segments with chyron titles, throw some color bars at the beginning, and we were done. Somebody would type up the labels on a Selectric typewriter and send the tape out to the client.

When we needed quick-and-dirty versions, I would disappear into the video closet, which contained two big first-generation VHS decks, made by Panasonic, with piano key control buttons. I would put together compilation reels, with one deck for playback and the other for recording, using the "Pause" buttons to edit. Those VHS machines sat in the video closet with a small Sony Monitor and, the most intriguing item of all, a Sony U-Matic video tape recorder with a one-tube camera. We also had some bright lights, lavalier microphones, and a tripod.

We used this gear for "media training" sessions. My boss would preside over these, and they were treated as solemn and important affairs. A spokesperson would come in and we would conduct practice interviews with them, then watch the results, critique the spokesperson and give them tips. They would practice over and over.

The idea was to get them comfortable in front of a camera, under bright lights, while being interviewed. They had to memorize different ways to mention the client's name and message. I was the camera operator. I did not know the term "videographer" at the time. When we finished, I would file the tape away for future reference. If you had no interest in public relations, it was soul-killing work, but it was better than being an account executive, which involved real work, such as calling people all day and pitching stories.

I stuck with this for a while and I began to lock myself into that video closet and make experimental videos. I discovered video feedback, where you train the camera on the monitor and create an infinite video loop. I made crude cuts, using the pause buttons and both tape decks, of found footage and stuff that I shot in the closet. I would record the latest song from my band onto an audio track and have an instant video.

This was heady stuff for me at the time. I smuggled the U-Matic deck out of the building one weekend, in two large shopping bags and a backpack. My girlfriend and I and shot some footage at home, on the roof and in the stairwells. She dressed up and I got to play the auteur.

I saved my money and bought a professional nineteen-inch monitor, made by NEC, which cost seven hundred dollars at the time. I bought it on Broadway and 47th Street and dragged it down into the subway, through the tunnels, onto the train, up the stairs, and all the way home. The box was shredded by the time I got it home, but the unit survived. It had more inputs and outputs than any device I had ever owned, including stereo amplifiers. I hooked it up to my Manhattan Cable box and was in heaven.

Manhattan Cable TV was new at the time. The cable box had fake wood grain on it and a row of push buttons that you would use to change the channel. These were given letters of the alphabet. The only one that was memorable, and which every New Yorker referred to by its lettered name, was Channel J. It carried public access programming, most of it strange, low-budget, do-it-yourself, experimental television. Channel J was infamous for showing softcore porn at night.

Home videocassette recorders were becoming common. Most people bought VHS machines, but I bought a Betamax. The tapes were smaller than VHS tapes and the picture quality was better, but not by much. The VHS machines were cheaper, but I was stubborn and believed in Sony and their Betamax.

Home videocassette recorders were revolutionary. No longer did you need to stay home to catch something live on television. No longer did you have to wait for a show to be repeated. Now, you had control over what you watched and when you watched it. You did not have to plan your life around the TV schedule. You could watch a show at your own leisure. Hollywood was distraught, thinking they would lose money to people pirating their product. In the end, they made more money, not just once from the movie theaters, but twice, when people bought VHS tapes and again, when they upgraded to DVDs. You could tape something and then, as a bonus, fast forward through the commercials. This infuriated the advertisers.

I had fallen into heavy drugs and spent as much time nodding out as I did awake at the desk in our bedroom. This desk was filled with clutter, so I kept a big ashtray above it on a shelf, where I left cigarettes to burn. They would fall while I was passed out and the burning butt would land on the cable box and fuse the plastic buttons to the unit. I had to pound on the buttons to get them to work again. I was too far gone to think about moving the ashtray or the box. The buttons continued to melt.

All of our money went to drugs, so the electricity and cable companies kept shutting off the service until one of us trudged over to pay the bills in person with a money order. I crept into the hallway in the middle of the night, stood on a stool, and screwed a splitter into the cable to tap into my neighbor's signal. I would creep out again at 5:00 or 6:00 a.m. to restore the cables. It was a horrible way to live and it got worse.

At work, I convinced the boss that we could do more of the video editing in-house, so we bought a two-deck three-quarter inch editing system with a dedicated control unit. I could now make clean cuts at the frame level. You would pause the tape where you wanted the "out" point, then fine-tune it with a rotary switch, scrubbing through each frame of the video. Then, you would do the same thing with the "in" point on the other tape. Finally, you would push a button and the machine made a clean edit. The finished tapes from this setup looked great. I had proposed the setup, set it up and learned how to use it. This was my moment of triumph.

The agency moved me to another floor and gave me my own room. I turned a desk away from the wall, lined the space at the bottom with videotape boxes, and curled up to sleep under my desk every day. I heard them come in, look around, say "where is he?" and then leave. I was declining as an employee and a human. The band was rehearsing and playing occasional gigs at crappy downtown clubs, but that was not going well, either. The PR agency hired a new guy, whom I trained, and then they fired me.

I found work at another agency as an account executive. They gave me a Selectric typewriter and my own Rolodex. A Rolodex was the analog version of a contacts database. Some of them were flat and had a smoky plastic, hinged lid that flipped up. The classic ones were round, with a knob on the side that you could spin to reach the card quickly. Each one held many small paper cards. On these, you wrote or typed your business contacts. You were as successful as the contacts in your Rolodex allowed you to be.

Rolodexes were ubiquitous and there was an etiquette to them. You would never look into somebody else's Rolodex unless you had permission. If someone was fired, you might inherit their Rolodex, but they usually took it with them. One other popular accessory in the mid-1980s was the Filofax, a leather book that contained your addresses, calendar and notes. It was sold as a system - on the razor blade model - every year you had to buy a refill. These office tools were later made obsolete by smartphones and computers. I was fired from my last public relations job for doing nothing. They always said something like "We like you, but you're not producing anything."

Using and copping drugs was now my full-time job. I asked my girlfriend to marry me. Things got much worse over the next three or four years. I sold everything I owned, bit by bit, for money to buy drugs. I got rid of the large record collection I had acquired over the previous fifteen years. I sold it, ten records at a time, to Sounds, a record store on the Lower East

Side. I would lug the LPs over there on foot to save bus fare and sell them for pennies on the dollar, all to support a heroin habit.

I sold my power drill and jigsaw. The routine at "We Buy," a pawn shop on 3rd Avenue and 11th Street, went as follows: I would walk in and show one of the old, sullen guys what I had. He would ask how much I wanted for it and I would timidly put out a number, usually low and coming from a desperate place. He would then hold the item up to the guy across the aisle and say "Guess how much he wants for THIS?" They would both laugh, then he would lowball me and I would always accept it. The script never varied. I was beaten-down and they knew it.

It was according to this script that I eventually sold my original Walkman, which I had proudly bought for two hundred dollars, six years earlier, for seven dollars. This amount would not buy a single bag of dope. I took the money and walked east to Avenue D, where I hoped I could beg someone for a bag at a discount.

My father had given me an Olympus OM-1, a fantastic manual SLR camera. There were no chips in it, no automated functions, with only a tiny battery for the internal light meter. I had one 50mm lens. To take a photo, you had to set the aperture and the shutter speed. You advanced the film with the thumb lever. Then, you would focus the lens and press the shutter release button. I did not sell that camera. I pawned it. As with so many other things at the time, I expected to get it back soon. I left it at a pawn shop in Hell's Kitchen in 1984 with a full roll of exposed film in it and never retrieved it.

The band dissolved by now, as well. I could no longer show up, the drummer took off on a long trip, and Roland must have felt abandoned. My wife and I were fading on drugs and I was ready to go down with the ship. I had no idea what was happening in the world of technology from 1984 until 1989. I was clueless when it came to computers and video games. I was more interested in how to pawn my typewriter. From 1985 until 1989, there are no photos. One day, I closed the guitar case and stuck it in the closet. I did not write a word for years.

32

BAGELMAN

I RARELY ANSWERED THE PHONE AND IT WAS PAINFUL TO hear it ring. There were no good phone calls by then, anyway, only creditors and the occasional call from my parents to see what I was doing. I had no way to muffle the ringer, so I would pile pillows and blankets on top of the phone. I never thought to unplug it from the wall.

We bought an answering machine at some point in the mid-1980s. It was a Panasonic and held two cassette tapes. The first was a minute long and held the outgoing message. The second tape was a regular 90-minute cassette and you could hear the message live as somebody left it on the machine. Being able to monitor calls was both gratifying and horrifying. Gratifying because we no longer wanted to talk with people, horrifying because it was always bad news on the other end of the line. The calls piled up and listening to them became one more awful chore.

I had not sold or pawned my Betamax machine yet, so I would sit around and compulsively watch war documentaries that I had on videotape. "The World at War" was my favorite, and I would watch all 23 episodes over and over. Sir Laurence Olivier narrated the entire series, which was a magisterial recounting of World War II. My wife would hear the opening music from the other room, where she was lying in her own misery, and she would scream "NOOOO." The music meant I would be catatonic in front of the screen, watching scenes from the Bataan Death March or the war in the North Atlantic.

I ran a scam with the local D'Agostino supermarket. Supermarkets did not accept credit or debit cards at the time. They were not wired for it. I discovered that you could pay for something at this branch by check, with a photo I.D., and tack on a twenty-dollar cash-back amount. I did this for twenty days straight. There was never money in my account, but I was not acting rationally. On the twenty-first day, I walked in and someone in the office platform above the cashiers shouted out "That's him!" I turned around and ran and did not return to that store for years. The bank dropped me, but not before my most interesting bank card scam.

In the basement of the World Trade Center was a foreign bank with futuristic-looking ATMs. These ATMs had the peculiar characteristic of dispensing up to two hundred dollars, whether or not there was money in your account at a different bank. They must not have been networked. As I recall, this worked only at night or on the weekends. I did it four or five times before they shut down my bank account.

In 1988, I got a job, via the Village Voice want ads, as a night manager at a bagel bakery uptown. They trained me, along with a few other young guys, all recent college graduates, how to manage the store. It may not have been a career path, but it was not the worst job. I hated working in the store, though, and being around people. After a month or two, they needed someone to fill in one night on a route to drive one of the trucks. I had already gone out on a route as part of my managerial training, so I volunteered.

I loved roaming the city at night alone in a truck. After a week or two, I convinced them to let me start driving one of the routes. After another few weeks, I had become a master delivery man and was able to bang out the route in little time. I could use the rest of the night to cop drugs. On Avenue D, the street corner dealers would see me drive up in the truck and they would shout "Bagelman! Bagelman! What do you need?"

There was a payphone right outside the door to this bagel joint on the Upper West Side. It was a partially enclosed booth. Most street phone booths had disappeared by the 1980s, partly due to people using them as toilets. They were replaced with half booths, where you stood, half protected from the weather and the ambient noise. Those standing half-booths were still better than nothing. You could almost get out of the rain. The sounds of the city were reduced a bit if you leaned forward. You could also use the booths as cover for smoking a quick hit of weed or snorting some coke or heroin. In my twenties, many of my life's most desperate moments happened at these phone booths.

When I copped drugs on Avenue D and the police presence was heavy, I would stand at the phone booth on the east side of D, at 9th Street, and pretend to talk on the phone. I wanted to look like I belonged there and was not just some guy copping drugs. The phone company programmed phones to make a loud noise when they were off the hook. If the phone was off the receiver cradle for more than a minute, an alarm would beep from the speaker to catch your attention and get you to replace the handset. This was useful at home, where you did not want to have a busy signal.

My solution was to wedge a nickel into the top of the switch hook, tricking the phone into being unused, even though I was holding the receiver to my head. I would stand there and act out the other half of many fake conversations, doing my best to impersonate a person with a real life. I would wait for a drug dealer to return or simply scout the street for trouble. People would call into payphones, though, hoping someone would pick up. I was often standing there, pretending to be talking on the phone, when it would loudly ring. This looked bad and I eventually had to stop using this subterfuge.

One sultry night in the late 1980s, at two or three in the morning, I was smoking a cigarette outside the bagel store when the nearby payphone rang. I picked it up and was propositioned by some local guy. I hung up, creeped out. There was no way to know whether he was watching me from a window or just trolling, calling local numbers. The phone company later made it difficult to call into pay phones from the outside, but it made for a strange subculture while it lasted.

33

MOTOROLA BRAVO

The bagel place issued me a beeper, my first piece of wireless technology. It was a Motorola Bravo pager, but we called them "beepers." Pagers had been around for a couple of decades, but they had been used mainly by doctors and other professionals. They were not the device you would expect to see on a low-level worker, such as a delivery driver. As with other high technology, pagers eventually trickled down. One of the sacred tenets of my generation is the idea that all great technology came from either NASA and the Apollo Program or the Department of Defense's DARPA - the Defense Advanced Research Projects Agency. There is a large grain of truth to that notion.

The pager was first patented in 1949, and it was used initially in New York City's Jewish Hospital. They were a reliable way to contact medical and emergency workers. They may have originally been only for professionals, but they entered the mainstream by the 1980s.

Mine was a sturdy rectangular plastic box with an LCD screen that displayed gray characters on a greenish-yellow display. It felt sleek and solid and was about half the size of a pack of cigarettes. It took one AA battery and had an on/off switch on the side. The locking battery door was on the bottom. You could press the on/off switch in, as opposed to sliding it up or down, to backlight the LCD screen.

There were two buttons on the top face. One was for reading and paging through the numbers you received. The other was for navigating the menus. The back of the beeper had a plastic belt clip that you would clip through your belt or directly onto your pants. You could set it so that it would make a noise or only vibrate. It was my introduction to Pavlovian conditioning by small, personal electronic devices.

The bagel bakery gave me this device so they could get in touch with me while I was on the nightly route. The boss would dial my number, hear a message, and punch in his number. The number would appear on the unit that was clipped to my white baker pants. I was expected to call him back as soon as I could find a phone. I usually ignored him. When I got back to the store, he would always say "What happened?" and I would put on my best blank face and reply with "What do you mean?"

He would complain that he had been beeping me for hours, to which I would say "I must have been in a sub-basement." One day, I said "There must have been satellite trouble." I thought that was impressive. How can anyone argue with that? I didn't even know if satellites were involved in the process at all. The manager just scowled at me and shook his head.

My mother used the beeper to try to track me down. My wife would desperately beep me when I was out copping drugs. She had lost her job and would sit at home, frantically trying to reach me. The number "9" was our code to indicate that it was important to call back immediately. I got beeps from her that ended with many "9's." The end was near for that job and for my habit.

I landed in rehab, where there were no phones, no television and no other diversions. There was an old wooden phone booth that we were allowed to use on a limited basis, but only to receive calls. The place was filled with union guys from New England who were trying to save their jobs. One of my roommates, a lineman for New England Bell, had a small cassette recorder and a Merle Haggard tape with the song "Sam Hill" on it.

I had never heard this song before but, in my diminished, impressionable state, it made an impact on me. I played it, rewound it, and played it again dozens of times. Music was one of the first things to reach me and make me feel alive again after so many years of living death. Twenty-eight days passed and I got out of rehab. I picked up drugs again the next day.

After a few months of deadly relapsing, I finally got clean and sober. I carried quarters around so I could call sober friends from payphones. I needed to find a job. I hated the corporate world, with its cubicles and what felt like fluorescent-lit slavery and kowtowing to people I did not respect. At the same time, if I had gotten a corporate job, I would have failed at it. My critical functions were impaired from the drug use, so I did what I was talented at and what I enjoyed - I found work driving a pastry truck. I was driving for a high-end patisserie in the West Village on a day shift.

My boss would not give me a beeper, so I found a service provider in midtown Manhattan and bought one for myself. At that point, I wanted one so I could stay connected to people and society. It made me feel part of the greater world again. It was 1989 and I was able to get a beeper with a Manhattan, 212, area code. This was before you needed to dial an area code to make calls within Manhattan. The island was just big enough to contain all of its numbers in seven digits, but this would not last long.

I lost my beeper's 212 number after a year, in 1990. The beeper company told me that Manhattan was running out of 212 numbers and I would have to switch to the new 917 area code. I fought to retain my original number. It did not take a genius to appreciate the value or cachet of a 212 number. "212" meant "Manhattan," it meant old New York, it meant everything that I identified with, especially as a transplant from the outer reaches of Yonkers, an immigrant from The Sprawl, 14 miles from the center of the universe. I had no choice in the end. We are always at the mercy of the telecommunications companies. I took the 917 number and sucked it up.

The beeper service provider was located on the upper floor of an old building in midtown. It was more like an office than a store. The sales rep gave me a complimentary, brushed brass beeper chain. The little alligator clip had a Motorola "M" logo etched into it, attached by a thin, six-inch gold chain to a tiny clasp that hooked onto a security hook on the beeper. When you wore it, the chain hung in a half circle from your belt loop to the beeper.

It was more blingy than I might have opted for, but I enjoyed wearing it. When you used a beeper, you would leave it attached to your pants and you could cycle through the functions by looking down and using one hand. This chain may have saved my beeper from falling off once or twice. It was not much help, though, when I was mugged the following year.

Two kids got me on a downtown "1" train. I was alone in the second-to-last car early on a Saturday evening. As soon as they came through the doors between cars, I knew something was up, but I did the New York thing and just stared straight ahead. They walked down the car, one sat next to me and quickly put me in a headlock, then threw me to the floor. The other piled on and they started to rifle through my pockets as the train slowed down and pulled into the World Trade Center station. They could not get the wallet out of my rear pocket – they were nervous and moving quickly – but they did spot my beeper, attached by my trusty beeper chain. One of them yanked it off, pulling the chain with it, and the two started to argue over who would get it.

The train pulled into the station and they jumped up, screamed "don't move or we'll blow your head off," then ran off. I staggered off at the next stop, got in touch with the transit police, then went to a payphone on Rector Street, where the only move I could think of was to dial my own beeper and punch in a "666," hoping it would spook them. I replaced that beeper the next day, even though it hurt to do it on my truck driver's salary.

1990 was also when I saw my first cellphone. It was a big, chunky thing from Motorola in the hand of a wealthy shoe salesman I knew. He loved standing on a corner in the West Village, talking on his phone. The idea of talking to someone, untethered on the street, was still a bit Dick Tracy to most of us in the early 1990s. It was gaudy, as well, and it felt like he was showing off. His phone bills were legendary, according to his secretary. A cellphone was a luxury item at that time.

Nobody was in the habit of talking and moving around. If you were away from home, you used a payphone. Cellular phones were originally called "mobile phones," back before cellular networks existed. They were radiotelephones and they operated on radio frequencies. Cellular networks involve a series of fixed cell sites, or towers. They form the network that your phone connects to. The technical magic is in your call transferring seamlessly from cell to cell, from one tower to the next as you move around. Cell phones supplanted all previous forms of radiotelephones.

34

ALIVE AGAIN

I HAD GOTTEN CLEAN AND WAS JUST HAPPY TO HAVE survived. It felt as though I had fallen back to earth, still smoking around the edges. It also felt as though there was some brain damage, as I was monosyllabic for the first few months. I wanted to say things and was painfully aware of my inability to formulate articulate sentences. The good news is that neural circuitry can re-map itself and the brain can heal to a great degree. My memory would never be as it was before I smoked all that weed, but I was happy to be alive. It was as though I had gone away and returned and, when I did, I began to realize that the world had changed.

Most of the people I knew had been planning their professions, getting the proper education, and working their way up whatever career ladder they had chosen. I spent most of my 20s bottoming out and burning my life to the ground. When I was 30, I was in rebuilding mode. I had no clue how to use a computer, nor was I interested in getting one. I had missed out on a lot of popular culture, as well – television shows and music. I had never owned a video game console. My answering machine still used cassette tapes.

I had traveled little, had bad credit, and no good job references. I still had plenty of time in life to pursue a solid, corporate career path, but no inclination. I enjoyed driving a truck around New York City, delivering cakes and pastries. Nobody was on my back, I could play music and chain smoke all day, and I made friends at every stop along my route. This was refreshing after all of those isolated years living outside of society. I discovered that I was good with sales and customer service. Luckily for me, the bakery was paper based. I carried around a tiny address book, a pen and a small pad.

I had missed out on the Windows-versus-Macintosh battles. I did not know what a hard drive was, and I could not make my way around an operating system to save my life. I would eventually catch up, but it took a while before I even knew I was ignorant. I – and almost everyone else – still carried quarters for payphones. Once I was able to re-open a bank account, I wrote checks to pay bills. At first, I had to use money orders from the Post Office. My apartment contained one remote control unit and it was for the television set. I used my typewriter for official correspondence and bronze tokens when I took a bus or subway. It was not a primitive life, but I had definitely missed out on things while I was underground.

35

PHONE NUMBERS

IT WAS MY CURSE IN LIFE TO FALL INTO A JOB THAT required me to drive to Yonkers every other day. I had escaped from Yonkers when I was 18 and never went back, except to visit my parents. Now, I had to drive there all the time and haul baked cakes and frozen pastries downtown from the commercial kitchen. My boss's West African bakers had been banished there years earlier, during a business merger.

I sat in traffic all the way back to the glittering coffee shop in the West Village. I would be alone with the thoughts in my head and driving around the city all day. I listened to the truck radio, always the oldies station, WCBS-FM, 101.1 on the FM dial, and I would sometimes carry a small cassette boombox with me, along with a few cassettes. I would listen to the whole side of a tape. If I needed to call the shop, I would stop at a payphone or, more often than not, wait until I got to the Yonkers bakery and call from there. Sometimes, I would walk in and they would say "Call Kelly," my boss back in the city.

I drove a truck in Manhattan all day, every day, and would often take one day off every two weeks. When the time came for my first vacation, driving south seemed like a good idea. I was unfamiliar with the term "busman's holiday" until then. My wife and I borrowed my father's car and set out for Washington, D.C. When we got there, the government went on strike. I had either not been reading the news or was just oblivious. I had dim knowledge about the country's geography, even at the age of 29.

I thought we should check out Virginia Beach, which I imagined to be a sleepy fishing village on the Eastern Seaboard. It was nothing like I thought it was, so we pushed on to what I imagined would be the next sleepy little seaside village – Myrtle Beach, South Carolina. It was a good drive. I was chain smoking, listening to music, and having a blast. Myrtle Beach was not what I expected, either, so we kept going. Why not Florida? I knew some people from the bakery who were in Orlando at exactly the same time, so we steered towards what I thought was another oceanside town. I did not know that Orlando was landlocked. I had assumed that everyplace in Florida was on the ocean. I still had a blast. We drove back gonzo style, in two long days, and I discovered that I loved road trips.

Shortly after that trip, my marriage fell apart. My wife kept the old phone number and I moved ten blocks south, still in the West Village of New York City. I got a new 212 number in the same 807 exchange that I had while married. When I got my own telephone number, it was as significant to me as getting a new bank account number, which I also did at that time. When I got my own beeper number, it felt personal. I carried it on me, clipped to my jeans.

The term "exchange" dates back to when the phone company labeled numbers according to the circuit they were connected to. Each circuit was housed in a central office. In a big city, an office might contain several exchanges. People don't use the term "exchange" anymore. It has become archaic. They don't identify numbers by the old exchange mnemonics anymore, either, but my mom and everyone from her generation did.

Our number in Yonkers began with 779, but I learned it as "SPencer 9." Roland's number was 337, or "DEerfield 7." My grandparents were at 237, but we said "BEverly 7" or, usually, "BE7." The phone company standardized them in 1955 and published a list of recommended names. If your number began with the numerals "58," for example, Ma Bell's list recommended that you use the names "JUpiter," "JUno," "JUstice," LUdlow," or "LUther" to remember them by. Our "SPencer-9" was not on the list, so it may have originated earlier or there may have been another reason, lost to the ages.

Up through the 1990s, you had a number and it was your "phone number," not your "home number." Some people also had an "office number" or "work number," but that was the extent of it. Having a fax number in your home was rare. You lived with your phone number until you moved somewhere else, at which point you got a new number and let everybody know what it was, just as you would let people know about a new mailing address. Somebody else would eventually be assigned your old number and they might have to field calls from old friends of yours who hadn't gotten the news.

When friends changed numbers, I would cross out the old one and write the new one in my desktop address book and the small, pocket-sized address book I carried around. You could track somebody's life and even their stability by how many cross-outs were beside their name in your address book. I might have come across as fairly stable if you were to judge me strictly by the longevity of my telephone number.

The phone company had automated the process of changing numbers, so you could arrange for them to deliver a message saying "the number you are calling has been changed. The NEW number is..." and their robotic voice would put more of an inflection on the word "new." This courtesy was good only if you were moving, not if two people were splitting up and one kept the old number. In the future, I would cycle through a few numbers before finally settling on one.

In the new apartment, I had a touch-tone phone. My first had been when I was married, in the previous apartment, but we had a rotary-dial phone up until 1990. The phone company used to charge extra for touch-tone service. They did this long after most people had switched over from rotary dial service to touch tone. It was another way to generate steady revenue, similar to the "service line repair" charge of a few dollars that they tacked onto the bill every month.

In the early 1990s, the phone company began to require Manhattan residents to dial "1-212" before dialing a local number. Until then, seven digits had been enough to connect you with your party. With the explosion of fax lines and beeper numbers, we ran out of 212 numbers. At the same time, the other boroughs picked up their own area codes. It became a technical necessity to dial the area code before the number any time you made a phone call. I was not happy about this. Aside from the practical aspect of it now taking longer to dial a number, there was a feeling that we were not living in what felt like a small town anymore. Dialing a local number with an area code felt less personal.

36

MACINTOSH

By the mid-1990s, I was managing the bakery's commercial department. It was better money than driving, but I missed being in the truck. This job was heavy on phone work. We took orders by hand and, even though we used abbreviations, it was a lot of writing. I was grateful to know cursive. I could not have functioned in that job without a mastery of clear, fast handwriting.

We made high-end wedding cakes and part of my job was to sit down with prospective brides and their moms every weekend. I had to explain wedding cake options to them and write out the orders by hand. All of the order sheets were in triplicate on carbonless paper. When I wasn't boxing cakes or haggling with bakers and cake decorators, I sat at a small desk, phone receiver glued to my ear, writing madly. We had a 4-line electronic phone system, but no computer. Even by the mid-1990s, most small businesses had no computer, but that was changing fast.

I had been only dimly aware of personal computing to this point in my life. The 1980s were a lost decade for me and I had spent the first few years of the 1990s putting myself back together. I had been obsessed with my beeper, but others were consumed with personal digital assistants. In 1992, when John Sculley was CEO of Apple, he described the Apple Newton as a "PDA," or "personal digital assistant." It featured a touch screen, handwriting recognition, and ran the Newton Operating System. It was not successful and had a lifespan of five years, which was about how long it had taken to develop it.

When Steve Jobs returned to Apple, he killed the Newton. There were many other PDAs on the market. The Psion "Organizer" had preceded the Newton, coming out in 1984. The IBM Simon came out in 1994 and was the first to combine telephony with a PDA. Nokia released the 9000 Communicator in 1996 and Palm began releasing PDAs in March of that year. The Palm Pilot was aimed at business and enterprise clients and it had diehard fans.

I ate dinner every week with the same group of people and would find myself lost when the conversation turned to computer hardware. I had always been mechanically and technically adept, but I felt like a moron when they began to talk about computers. I would sit quietly until the conversation shifted to other subjects. They would drop words like "megahertz," "SCSI" or "hard drive." These people were gearheads, so they got worked up during the discussions.

One of them invited me to his apartment to show me his new Gateway 2000 computer, which he switched on with a flourish. A tiny video played on the screen and he cackled with glee at the machine's ability to play video. I still did not know what you could do with a personal computer. I knew that he wrote poetry on his. Another friend used his Mac to create music and earn a living. He had turned his whole living room into a studio, centered around a large computer monitor. Still another friend was a 3-D artist who absolutely loved her Amiga computers.

In 1994, I was at my parents' house for Christmas and my brother, a successful writer, told us he had recently bought an Apple PowerBook 150. That got my attention. He raved about it and explained how it had made his work easier. I was beginning to wake up to what was happening around me. At the all-night magazine stands I used to visit, I noticed that the section devoted to computer magazines was getting bigger and bigger. I began to browse those magazines, but I still did not think I needed a computer.

I was also losing my mind at this time, having finally broken up with the woman I had been having an affair with. My marriage was over and this relationship, built on a horrible foundation and teetering for over a year, finally ended. I found myself alone for the first time in 13 years. Whatever other steps I took to find my way out of this – from walking the city streets at night, to riding my bicycle, to trying to find love via the Village Voice personal ads - the real healing came from an old habit. I began to write again.

I did not plan it and I did not produce anything great, but some deep instinct made me stand on a milk crate in front of the closet, pull my Smith-Corona electric down from the top shelf, put a piece of paper in the roller and start typing. I needed to get stuff down on paper. It was desperate, diaristic stuff, but it didn't matter. The main thing is that I was doing it every night, pounding out page after page. I wasn't reading it, just typing like mad. When I finished one page, I would lay it face down to the left of the machine and start another one.

My neighbors were an old couple who had been in the building since the late 1940s – Frank and Vera. He was couch-ridden most of the time. He was a veteran and still displayed a framed letter from the army on a china cabinet. Vera let me read it. He had done intelligence work in the China-Burma-India Theater in World War II. It took some imagination to reconcile the description of his heroics, along with the photo of him in uniform at a jungle camp in 1945, with the retired mail carrier on the couch next door to me. Frank was done, played out, floating down to the runway on fumes. She was the more animated one. She would scream odd things at all hours.

The walls in that place were thin, even though it was an old building, a "pre-war building," as they call it in New York City. I would be in bed late at night and be startled by a loud sneeze in the next apartment. Jacked up on tea and coffee at 3:00 a.m., I was hyper-aware of my typewriter's noise. I would stop typing and be left with the hum of the motor, then start typing again, the keys smashing into the paper like little jack-hammers. I was curious about computers by now, but I was also rcsistant. I did know that a computer keyboard was quieter than a typewriter. This was the germ of me finally buying a computer.

The manager at the patisserie convinced the boss to buy a computer, mainly on the pitch that he could use it for book-keeping, but the real draw for the boss was that he could create signs. He wanted a toy as much as anything else. Many people who bought computers then, after the pioneers and the early adopters, simply had some disposable income and were curious.

He bought an IBM machine, a PC, and proceeded to make up a bunch of signs, 8.5 by 11 inch masterpieces that said things like "Please Form A Line to the Left" and other banalities that you find in a working bakery. Hard drive space was in short supply and he was so concerned about it that he deleted the text files every week. He forbade anybody but himself or the bookkeeper from touching the machine. For this, of course, he made the appropriate sign.

I went in to look at this marvel the first week he got it and saw first-hand what you could do with a personal computer. It was a basic Windows machine, but it was a whole lot friendlier than the previous personal computer I had seen in the mid-1980s, the IBM 5150, at the PR agency where I had worked. I ignored that first computer, then forgot that computers existed for the next ten years, until I was finally clean and sober and conscious again.

A computer was a serious investment and I needed to know what I was doing. I began to ask friends with computers which one I should buy, a PC or a Mac. I didn't even know the difference between them, so my initial question was usually "I want to buy a computer, but I don't know what to do. Can you help me?" I began to learn a few things. There was no internet for me to jump on and do research. Most people I knew did not have a personal computer. The ones who did used it professionally or artistically.

I began to notice a pattern. The people who owned PCs, the machines that ran Microsoft Windows, became dismissive and even nasty when they talked to me about Macs. They defended PCs with venom. Some of them insulted Apple and were dismissive of Macintosh computers. The Apple people, on the other hand, seemed tolerant. Most would listen to what I wanted in a machine, they would tell me what they liked about theirs, and then they would usually say something accommodating and tell me to get either one. This attitude is what swayed me. The PC people were rigid and often hostile. I respected the Mac people more.

It wasn't a question of Macs being cool, either. Apple, at that time, was not a cool proposition. The company was on the rocks. Steve Jobs was long gone. Gil Amelio was running the show. They had not come out with an exciting product in a long time and the news magazines were saying that Apple was close to bankruptcy. People were hoping Dell would buy them.

I was about to spend over two thousand dollars, a big percentage of my yearly salary. Did I want to throw in my lot with a loser company? One friend told me he would help me out if I got a Mac and that gesture of kindness, as much as anything else, helped me to decide. I didn't think about the long-term ramifications. It was partly an emotional decision, but I also liked the way the Mac interface worked, from my limited exposure to it. As with my earlier decision to buy into Betamax, I was willing to risk hitching myself to a loser again.

I found an authorized dealer in Westchester. I don't remember how I found them, probably through an ad in the New York Times or the Village Voice. I did my research, got the money together and ordered a Macintosh Performa 5115C, which had a 60-megahertz PowerPC processor. I had the dealer add sixteen megabytes of extra RAM, for a total of twenty-four megabytes. It had a 350-megabyte hard drive, a floppy drive and, at the cutting edge of technology, an optical CD-ROM drive.

The machine shipped with a 15-inch color monitor and an external 14.4k modem, which I assumed I would be using for sending faxes. People called this design the "pizza box Mac" because it was thin and square. I felt as though I were consulting with wizards and that I needed to trust them absolutely, as I knew almost nothing.

37

HELLO WORLD

I WAS WORKING SIX AND SEVEN DAYS A WEEK, SO I HAD this company deliver my new Apple to the bakery, which they did. On the big day, some guy drove up in an old station wagon and made me sign a receipt. A tall box held the computer, the monitor, the modem, the manual and the media on compact discs. A life-sized color photo of the machine was printed on the outside of the box.

When I was done with my shift, I borrowed one of the refrigerated pastry trucks, put the computer in the back, and drove seven blocks to my building, where I dragged it gently through the hallway and up the stairs. I returned the truck and went to the Hudson Diner to grab a bite before I came home to tackle the assembly. It felt like an important day.

I opened the manual and began to read, from the initial eye-glazing FCC statement, to the Table of Contents, to the "Welcome to Macintosh" statement, which I took to heart. I felt as though I were being personally welcomed into a new community, a special tribe. I took it all in. I read every word slowly. I was impressed by the simplicity of the directions. They dissolved some of my fear. I followed the instructions and hooked up the monitor, the keyboard and the mouse, using the ADB and monitor cables.

The monitor looked like a beige TV set. The computer was as big as a pizza box, but thicker, and heavier than I expected. The modem was a small tan-colored box the size of a paper-back book. I sat the computer and monitor on top of a steel platform with a sliding keyboard drawer. I had bought this ahead of time at an office supply store. I plugged the computer into a power strip and left the fax/modem in the box for the moment.

I took a deep breath and flipped the page to the section titled "Turning the Computer On." I had never turned a computer on before, so I was excited. Nothing happened for what seemed like a long time, then I experienced that magical moment, something which surely had been planned and discussed in Cupertino, when I heard a chime and was greeted with a smiling Mac face at the beginning of the startup process. This was followed by the march of icons for the system extensions as they loaded across the bottom of the screen. It was March 27th, 1995.

I could not do much. Hooking it up and powering it on was a victory but, the more I toyed with it, the more I realized I had no idea what I was doing. There was no free ride. I would have to work to learn what was going on. I should not have worried. It's hard to overstate how big a hole there was in my life that needed to be filled at exactly this moment. An awful relationship was over, as was a troubled marriage. I felt lost. I should have been celebrating instead of mourning, but I didn't feel that way at the time.

I had been in a limbo state for months. I had too much time on my hands. I had been walking around the city or sitting at home and having long conversations with my cat. My life felt empty, without love or meaning. I was howling at the moon. I had begun to deal with the romantic loss by writing. The typewriter noise had led me to buy the computer. My salvation had just been delivered to me, but I didn't know it yet.

My obsessive nature could now be put to good use. My transformation into a computer geek helped me to regain my sanity and survive. I switched one obsession for another and figured out how to fix technical problems and to master software. This got me through the darkest days. It was my rebirth and the start of my evolution as a trouble-shooter. "If you can see the Macintosh desktop" said the manual, "skip now to 'What's Next?' later in this chapter."

I skipped to "Learning the Basics" and went through the section on how to use the mouse, then did the initial tutorial, which showed me how to navigate the desktop. The manual went no further, although it did point me to other resources, such as the "Guide menu," "other manuals" or "Apple's customer support hotline." Nowhere were the "internet" or the "world wide web" mentioned. They existed, but there was little useful information online at this time. I had to figure out how to create and save documents. I used other manuals and I experimented. I called some people, but I was ashamed at the basic level of my questions. I documented everything in my journal.

My friend Louie, the biggest computer geek I knew, came over with something he was clearly proud of, a one-gigabyte external hard drive, along with a fat SCSI cable to hook it up to my machine. I learned that "SCSI" stood for "small computer system interface" and it was pronounced "scuzzy," which made me happy in a silly way. The plug was long and had a bunch of tiny pins in it. It also had two set screws on it to prevent it from pulling out. You pushed it into the corresponding female jack in the computer, tightened the set screws and had to restart the machine.

An icon would appear on your desktop and this was a representation of the hard drive. It was called a "volume" and you would open it like a folder. It made immediate sense to me when I witnessed it. Louie gave me a ton of shareware and some other fun things, such as a "Flying Toasters" screen saver. I worried about images burning in on my screen. I was impressed with his knowledge of the machine and the operating system and realized how much I did not know and needed to learn. He finally left and I was alone with the machine.

That night, I worried about how I would learn to use Photoshop. I had version 2.0 and it was pretty simple, but I had no skills and was lost at first. It would take some time for me to master Photoshop. In the meantime, I would play with the menus and, if I were able to make something happen, it was a victory. I hated asking people for help, so I would fumble around until I figured something out. I was becoming better at the word processing program, WordPerfect, every day. I don't recall why I chose WordPerfect over Microsoft Word, but it was not my smartest move. Years later, when WordPerfect was discontinued and no longer worked on a modern Mac, I would pay the price in some all-night conversion sessions.

The simplest and most satisfying thing I did in those first few days was to play with fonts. It may seem trivial now but, until personal computers came along, it was difficult to create a document with different typefaces. With Letraset or Prestype, some patience and a good eye, you could lay out documents. Aside from that, it was not possible for most people to do so at home. I was ten years late to this party. To select some text, then change the font, size and style in a few seconds felt like magic.

That first week, I met the cyber world of my desktop and fell in love with it. I was about to meet the internet. The crew of people I hung out with began to talk about being "online." I was curious about the Global Village 14,400 kilobyte-per-second modem that came with my computer. Apple clearly expected me to do something with it so, within a few days of setting up the computer, I hooked up the modem. It plugged into the computer with a rounded 9-pin "Geoport" plug. I hooked a phone line into one of the jacks on the modem, installed the Global Village software and sent a fax to my parents. They were the only people I knew who had a fax machine. It read "Happy Birthday Mom" and I used a dingbat character. Dingbats was a symbol font that let me create simple graphics. They were a lineal progenitor of emojis.

Within a few days of getting this computer, I experienced my first crash. System 7, the operating system it shipped with, was the latest version in the line of operating systems that ran Apple computers. It was not stable. I started the machine one day and everything became garbled on the screen, which then froze. I panicked. I wanted to fix it, but I was worried I might break something. I would discover later that computers were tougher to break than I thought, but I didn't know that at the time. I called Louie and left a desperate message, but he did not call me back. I figured he was either out of town or avoiding another needy call from me. In those pre-cellphone days, being out of town usually meant that you were unavailable by phone.

I have always had an obsessive nature, especially when it comes to mechanical things. By the time I was eight years old, I had taught myself how to disassemble my bicycles down to the ball bearings and reassemble them. I enjoyed being able to build and fix things around the house and to make car repairs. Feeling powerless over a disabled computer did not sit well with me, so I did the next most difficult thing to feeling powerless – I reached out for help. The New York Times printed a section of small ads from people who offered professional help. One ad was for a computer person who would come by your house. I called this guy, who professed to be a Mac expert and who seemed to recognize the distress in my voice. He came over, dressed in clothes from his day job.

He had the confident air of a successful surgeon. He sat down, admired the advanced technology of my new Mac – "OH, a CD drive! Twenty-four megabytes of RAM!" – and then he went to work. I watched him, but I did not learn much as he sorted things out. I was grateful and more than happy to fork over eighty bucks for the hour he was there. I don't recall making a vow to never let this happen to me again. Watching him work, though, lit a fire under me. I set out to learn how to get under the hood of my Mac, maintain it, and perform simple fixes. I kept his business card scotch-taped over my computer for years, though, just in case.

With Mac Operating Systems 7, 8 and 9, you had to "rebuild the desktop," "re-set the PRAM," troubleshoot preferences and run diagnostic utilities all the time if you wanted to remain trouble-free. Those machines were never truly trouble-free, though, and you might be stuck re-installing software for hours, especially if you were obsessive about it. I had nothing else to do but go to work and rue being single, so part of me welcomed long, late-night sessions swapping floppy disks and trying to figure out the answers to obscure technical questions.

38

NOMADCAT AND AMERICA ONLINE

My Mac came with a binder of CDs – the bundle of games and utilities that Apple was pushing at the time. Included was an AOL disk. It was around this time that everybody I knew began to receive promotional installer disks, both floppies and CDs, in the mail from AOL and CompuServe. I popped the AOL CD into the optical drive and ran the installer. I had only one phone line and I had not yet bought a splitter, so I unplugged my telephone and plugged the modem into the wall jack. I clicked a button and marveled at the modem sound.

I learned later that the distinctive whistling noise a modem makes is the sound of one modem talking to another, making the connection. The word "modem" comes from "MOdulation/DEModulation." I felt that familiar sensation from my youth of entering the space age. The modem noise was the sound of a connection to something bigger. My phone line would soon be busy all the time. This was true for other people, as well. If you were online and, like most people, had a single phone line, you could not talk on the phone.

AOL made the technical connection and then it asked for my information and a credit card number. It asked me to choose a username and I thought about it for a second, tried "Paul," which was taken, then decided to ditch my real name in favor of NOMADCAT. I had come from a long line of nomadic Greeks and I had a cat. Real imagination at work here, but I liked the name. The system accepted it and I took that as a sign from the technology gods.

I signed on, heard a computer voice say "Welcome" and was suddenly online. This was a new dimension. Initially, there was little to see. The thrill was in having gotten there, a non-physical place that felt real. I was excited at the idea of interacting with other people, even though I had not yet sent or received an email nor been in a chat room. I read some news, poked around the kiosk-like front page, and moved on. My first email has been lost to posterity, but I was busy online from that moment on. Within a week of turned on that computer for the first time, I was typing in my journal, between the entries about the bakery and my destroyed love life, that I was "exhausted, been up late a lot on AOL."

I developed an online life. As "NOMADCAT" on AOL, I joined online conversations, tried to get laid via AOL chat rooms, checked the weather, read astrological predictions, and wasted a lot of time. The main attraction of cyberspace in 1995 was in finding other people. Useful services and information were still sparse.

AOL chat rooms always moved fast. They were named and you could see how many members were in each room. It seems like a waste of time now and shameful to admit, but it was novel and exciting to converse with anonymous people online. These chats had an odd quality of rapidity and randomness. A "chat room" held up to 23 people at a time. Any person could be talking with one or more other people simultaneously. The text would fly by and you could half watch what other people said while conducting your own conversation. It was like running around in a crowded room and chatting with different people who were also running around at the same time.

I knew that people were living in their own universes over at CompuServe and E-World. They were in a different solar system. AOL was big enough to get lost in for the moment. The world wide web, which people in the media were trying to explain to the general population, was another matter, and AOL could not overlook it.

Articles appeared in the newspapers and stories aired on TV about "the Internet." The old media was reporting on the new media. The internet eventually became something that people wanted to experience firsthand. It picked up its own momentum. For many people, AOL was a gateway to the internet. For others, it was everything. AOL released a proprietary web browser at this time that may have been one of the worst pieces of software ever written.

My memory of that first AOL browser is hazy and I can find almost nothing about it online. There is so little information that I would think I dreamt up the whole thing if not for one clear and persistent mental snapshot of me staring into my monitor and wondering why I had tried it. Early in 1996, AOL was scrambling to offer web access and prevent customers from leaving their ecosystem. The game was over for them, but they did not know it yet. I don't remember whether I had to install their browser separately or whether it was bundled in that version of AOL. I believe it was a button or a menu item.

Either way, I clicked on a button that said "Internet" or "World Wide Web" and a browser-like interface launched that read "AOL Browser Version 1." My main memory of the AOL browser is that it did not work. It would not launch pages and it just froze there forever. It displayed broken image icons and jumbled formatting. AOL bought Netscape a year or two later. They tried to bolt on a search engine and another browser, but they also failed. Their reputation began to suffer.

In 1996, AOL was still operating under the premise that they could co-exist with the web and remain relevant. Their population was growing every day and they had millions of users. Apple's "eWorld" was failing and would die before the end of the year. CompuServe was in an Avis situation, advertising against its bigger competitor and struggling to remain viable.

AOL was an easy way for people to get online and test the waters. The installation and user experience were simple. The world was still dialup, so all you had to do was insert an AOL disk, make a few clicks, and you were online. That was their strength - ease of entry. An America Online installer disk became a ubiquitous piece of late twentieth century technological trash.

I had chosen AOL because it was easy and seemed to have critical momentum. I knew about CompuServe and had heard of eWorld, but I landed at AOL because their disk was in hand. It was no small thing to sign up with an online service. It cost money and you needed a credit card. AOL billed by usage. The longer you stayed online, the more it cost. They eventually moved to a flat rate model but, in the early days, you would read stories about people getting monthly bills for hundreds of dollars. The abiding law of capitalism – charge what the market will bear – was in effect.

At the same time, internet providers were starting to provide unlimited access for a set monthly fee. AOL was still charging by the hour. My friends were talking more about the internet and I knew it was free. I did not know what it was or why I might want to go there, but it was calling me. The main impediment was technical. I needed to figure out TCP/IP protocol and get my modem to talk with another modem without the help of AOL. I was quickly learning how to figure things out on my own.

By this time, I had also become a student of computer history, especially the history of the Macintosh. I had eaten up every book I could find on the subject and, on a trip to California in 1996, had made the pilgrimage to 1 Infinite Loop, in Cupertino, to visit the Apple Campus and buy a tee shirt in the gift shop. I drove up, saw the lawn sculptures of the desktop interface elements, and smiled. I bought some Apple swag. At the time, the Apple logo was still sporting the rainbow livery. I felt as though I were a part of something.

39

THE WEB

SOMEBODY RECOMMENDED A LOCAL PROVIDER, Panix.com, which stood for "Public Access Networks." There were only a few providers in New York and I went with Panix, although I thought that "phantom.com" sounded cooler. The people at Panix were helpful and, within a day or two, I had installed some software, configured my settings, downloaded the Eudora email client, the Netscape Navigator browser, and connected to the internet. It took some tweaking and tuning, but I was able to eat dinner with the crew the next night and proudly announce that I had gotten onto the web and now had an email address.

I don't recall much of what I browsed in my initial "web surfing," as people called it, when the web was a recreational activity for everyone but scientists and defense workers. There was not a lot of useful stuff online for most people. Some version of online banking had existed in the past, and electronic communications were not new. Email had existed in various forms since 1971 for researchers and scientists. The first website had gone live four years earlier, in 1991, and that first site contained information about the web. Most people online in the mid-1990s were simply looking around or trying to build websites of their own.

The number of people who used the internet before 1994 was miniscule compared to the flood that came after. My baptism into the online world came at the moment when it was opening up to the masses. I launched Netscape Navigator and looked around. I bookmarked things. I saved my bookmark files for years, as time capsules. The earliest file I have is from January of 1996, so it's what I had accumulated after eight months of web activity:

Panix - Public Access Networks Corporation

A Beginner's Guide to HTML

Library of Congress Indexes to WWW Services

Britannica's Lives

Panix Mac internet Pack

WebCrawler Searching

WebWeather

Astrology Horoscope Jonathan Cainer (US)

Horoscopes / Astrology: Weekly Cyber-Stars from John James

REAL ASTROLOGY

Psychic Advice Line Daily Horoscopes

Lycos, Inc. Home Page

WORD

Netscape: Astrology on the Net

ASTRO INSIGHTS by Bridgett Walther

Mystic Stars Home Page

USED KARMA Home Page

Astrology For the Week

Hermes' Foot

Mercury Mail

Astrology - Horoscopes - Jonathan Cainer's Zodiac Forecasts (UK)

The BIG List: Links

HTML ColorEditor for CGI

Pattern Land!

The History Channel

The Weather Channel ¨ Home Page

Today's Stars

Arizona Highways Online

JavaScript Authoring Guide

This list may say more about me than about the state of the web, but it's useful to look at. Nowhere is there a link to any commerce or banking. I was still paying bills and shopping in the real world. Amazon existed, but it was mainly for books and I did not start to buy from them immediately. eBay was around, as well. The first receipt that I saved, and it was probably not my first online purchase, was on June 24th, 1996, from software.net. I paid $49.99 for a version of Eudora Pro, the email application. Most of the receipts I saved at the time were for software. The deluge of book receipts from Amazon began months later for me.

There are no news sites in this list, but I know I was reading the New York Times site on a regular basis. There simply was not much useful stuff out there. Entertaining, yes, but useful, no. There is something glorious about that now. I found a lot of good information on web-authoring, technical support, and the weather. I saved a bunch of links on astrology. The web astrology people were pioneers in running sites that were updated daily. These were the precursors to blogs in many ways. I read them when I was in an emotional pit and looking for solace in the intangible.

My first search engine, WebCrawler, is listed. This was the first search engine to index entire web pages. It was a revelation. Search engines work by indexing web pages. They do this through "bots" and "crawlers," which are little programs that collect a website's content, links to other websites, and meta-data. They crawl the websites and store the information in a central place. Then, they index it to deliver the most relevant information first. Finally, they serve the results to the user who types in "Miter Joint Photo" or any other query. Websites change and grow constantly, so the process is continuous. Searches in the mid-1990s did not bring back many results, which is why so many early web sites had a section called "Links." You could learn a lot about a person by the list of links they posted.

Before search engines appeared, people posted web pages devoted to lists of other pages. This made life easy. If you could not click on a link, you had to enter the URL – or web address – by hand. These addresses contained slashes or tildes and, if you messed up one character, you would get an error.

You would hit a site, check it out, and then see what links it contained. Google later exploited this when designing their algorithms, but that was a few years in the future. Google did not exist yet, nor had I discovered Altavista, which would become my primary search engine for years. In 1996, it looked as though Netscape was the great force that would dominate the web. New releases of Netscape Navigator were big events.

The amount of information online was growing every day. The bigger issue was that people did not yet know what they could search for. There was no model for existing and functioning online. One phrase the media loved was "the information superhighway," which describes a place that you navigate and is full of information. It's not the best explanation of what actually happened online. The label did not last long. The public did not know how to think about this new space. It was difficult to explain to people what the internet was. Someone might say "What's it like? What do you do there?" It was easier to show than it was to explain it. Once you witnessed a web browser, you understood it immediately. The internet was not yet ubiquitous.

Nobody at this point was concerned with privacy issues, with being tracked, with their data being harvested and sold, or even with identity theft. These things would all come quickly, though. While most people were thrilled just to discover the online community, there were some opportunists who recognized that you could make money, one way or the other, on the internet. "Cookies" were little bits of code that a website would leave in your browser. They were invented in the mid 1990s as a way for a website to remember your preferences. People first began to hear about them in 1996.

In the mid 1990s, "Cyberspace" became another popular term, but it has faded. It always had too much of a sci-fi connotation for me. People began to put up simple web pages, partly as a challenge, partly to stake out one little corner of this new dimension. "Check out my website!" was another big phrase at the time. These websites usually had a photo, some pithy message, links to other sites, and a contact link. More often than not, there was a message, saying "Under Construction."

This idea of being under construction was so common that people began to think of clever or ironic new ways to say it. For my own website, I just settled on "Forever Under Construction." In reality, it was not. It was a static page for years, but that was fine with me. It was something I could show people, something I had built on my own by stealing HTML code from other sites. I had to learn about hosting and FTP clients, as well as a few other things, but I enjoyed the challenge. As with everything else computer-related, it helped me to pass the time and get over that awful relationship.

In the beginning, I was spending more time on AOL than on the web. That's where I found community. I wanted the social contact. AOL had chat rooms, message boards, news and information. The company gave it their best shot at trying to maintain a closed system. They were battling the world wide web, one of the most powerful forces humankind has ever unleashed. For the next year, I was using both AOL and the world wide web simultaneously.

The online world was a welcome escape from physical reality. You could read the news before it came out in print. You could find odd photos and information that you would never come across in your offline literary travels. You could meet new friends, contact old ones, be anonymous and interact with strangers, be crazy, hurl verbal bombs, and start what came to be called "flame wars" – long, crazy arguments online. And you could do it all while sitting at a desk in your underwear.

Being online was the gateway to another universe. My modem was connected to the phone jack in my living room by a 50-foot phone cable. I had carefully run it along a baseboard, up a wall, along the ceiling and down again, with small "U" nails. I would make the dialup connection over these copper wires, then wait as I listened to the wailing modem and the eventual connection. I experienced a Pavlovian response of relief and excitement when the connection was made. I would start my email program and go make coffee. It took a long time at dial-up speeds for my mail to arrive. This whole time, my phone line would be busy to the world. It was not uncommon to send someone an email, saying "I'm trying to call you."

I had been a magazine hound since I was old enough to read. My childhood house always had stacks of magazines lying around and I loved the format, the longer pieces, the glossy pages and the photos. My daily routine for years included hitting the well-stocked newspaper and magazine stores of the East Village – Gem Spa and a place on Avenue A near 4th Street. I spent a lot of money, happily, on history magazines, music magazines, aviation rags, literary magazines, eclectic journals and more. I would eagerly wait for the next month's issue of "Four Wheel Drive" or "Military History Quarterly," just as I had waited impatiently each month as a kid for the next issue of a favorite comic book.

In the mid-1980s, my dad began to give me copies of a magazine called "American Heritage of Invention & Technology," a magazine devoted to the history of technology. I cherished every issue, read them all from cover to cover, and kept them all on a shelf in my room, along with other magazines.

The web was different. You could not read it from cover to cover. Hypertext felt revolutionary at first – clicking on a link in the middle of an article – but I came to miss the familiarity of linear stories. On the web, you could never run dry of your favorite subject. The amount of content grew every day. Huge archives of old material began to accumulate online.

What seems intuitive now was brand new and unfamiliar then. I had to explain email to both my father and my brother, who had bought his PowerBook a year earlier, but used it strictly to write his novels. He had never been online and was a bit dubious about it. I had to describe it to them: "You send the email and then it's done. The other person doesn't get it until they sign in and check for it. No, you don't have to be online, waiting for it." Email was something that both my father and brother were interested in. I did not tell them about "spam," which is electronic junk mail. Early on, though, I started collecting spam with ridiculous subject headers: "Facelift in a Bottle," "Pack Your Thong, You're Going on A Trip," "Find Sexaholics in Your Area!" "End This Abusive Advertising" and many more.

40

MAIL

Writing a physical letter took more effort than sending an email. There was more forethought involved. It was something you might do on a whim, but a letter usually meant that you had something to say, even if it was just "I miss you." You needed to get a piece of paper and a pen or to put a sheet into your typewriter. You composed it, you signed it by hand, and then you folded it and put it in an envelope.

There were different techniques for folding and people were wedded to their own method. A legal-sized envelope was easy – you folded the sheet from the top and the bottom and then put it in the envelope. Smaller envelopes were different. I never gave it much thought until a teacher in grade school decided to take five minutes one day, on her own initiative, and show what she said was the proper way to put a standard sheet into a small envelope. She folded it once in half horizontally, then she folded the result from the sides, once from the left and once from the right. She stuck it into the envelope and said, "That's how it's done." I took it as useful knowledge and never forgot it.

You sealed the envelope. Self-adhesive envelopes were not common in the 1990s, so you had to use a sponge on the glue or lick it, being careful not to cut your tongue while doing so. Finally, you had to address it. Many people kept small return-address stickers around. Some had rubber stamps. Others just wrote the return address by hand. Your address always went in the upper left corner of the envelope. You could write it in the center of the envelope's rear side. If it was a personal letter and you were feeling playful, you might even write a tiny note on the rear of the envelope.

The recipient's address went in the center on the front, a bit to the right. There is a certain majesty, a mystique, about addressing an envelope to a specific geographic location, "P.O. Box 333, Grand Central Station, New York, NY" or any of the billions of addresses around the world. Writing an address on the envelope evokes a sense of emotional travel. We take for granted that this letter will find its way to a physical destination and be delivered by hand, but it's an amazing thing. Email has its uses and its own magic, but that sensation of physical travel is not one of them.

The proper postage went into the upper righthand corner. You had to know how much it cost to send a first-class letter and keep track of changing postal rates. There were no "Forever" stamps until 1995. Usually, you had an envelope filled with stamps sitting around and this might include what were called "makeup stamps," of one, two or three-cent stamps. You would use these when postage went up if you still had old stamps around. Either way, you had to lick the stamp and taste that special glue flavor. The first self-adhesive stamps appeared in the U.S. in 1974. They were issued ten years earlier in some tropical countries, where humidity killed the glue.

You would leave the letter on a table or someplace else until you went out, at which point you would stick it in your pocket or your bag. In your travels, you would pass a mailbox, where you would pull the metal door and get that reassuring clunk when it closed. If you were paranoid or careful, you might check to make sure it actually had fallen into the mailbox. If you lived near a post office, you might go inside and drop it in the chute. That was one way to make sure it went out as early as possible. If you cared about speed, you would read the sticker on the front of the mailbox to note the pickup times. After you mailed the letter, you walked away and moved on with your life. Your part was done for now.

A week or two later, you might receive a reply. You would see somebody's familiar handwriting on the envelope, you might note the stamp, then you would rip it open and read it. When you were done savoring this object, this combination of the physical, emotional and intellectual, you might fold up the piece of paper and toss it onto a pile, to save for the future, maybe to reread or to throw out one day with a bittersweet feeling. Or you might rip it up immediately.

After email appeared, people derisively began to call postal mail "snail mail," as though speed were the only thing that mattered with letters. Postal mail is valuable to our common memory. Paper seems to persist longer than digital bits and bytes. The future will tell, but we have correspondence from many centuries ago and much lost electronic mail from the past few years. Correspondence is one of the most important records that historians have. History is reconstructed from letters, journals and official documents.

Archiving email is difficult. Much of it is server-based, transient and fragile. If someone dies, somebody else has to get into their email and archive it or it will disappear. Nobody prints their emails out. Someone who keeps it locally, on the computer, accessible via an email program, has to deal with changing formats, dying programs, and reliable backups. It's easy for digital files to disappear, but you can toss paper letters aside, forget them for years and, barring a fire, they will still be there one day. If you're rigorous about backing up your computer and if you don't have a catastrophic crash, your stuff might survive for as long as you do. It's not a sure thing.

41

REDUNDANT, MULTIPLE BACKUPS

I BOUGHT MY FIRST ZIP DRIVE IN 1995. I WAS WORRIED about losing my files and information. I knew the adage about there being only two kinds of computer users: those who have already had a catastrophic crash and those who were about to have one. Floppy drives, which my first computer shipped with, were limited. Floppy disks were slow and small. Each one held just a few megabytes of data. They had been around for a long time. When I was a Wang temp worker, I stored some personal files on a five-inch floppy disk. IBM, Atari and Apple had been using floppy disks since the 1970s.

They were standardized at 3.5 inches by the time I got into computing. The online companies and virus protection people were still sending floppies in the mail for promotions. The CD, or "compact disc," eventually replaced them, but the installer disks for the operating system on my first computer were floppy disks. The terminology is to call optical media "disc" and magnetic media "disk." I don't know how this naming system evolved and I doubt that most people know or care about the difference.

Reinstalling the operating system on that first computer consisted of shuffling 10 floppy disks in and out of the machine, over and over. Messages would pop up on the screen: "Insert Disk 3" or "Re-insert Disk 1." The floppy drive would whir, grind and click. The computer would slowly suck in all the information, install it, spit out the final disk and you would be done. It was a tedious routine and it took forever. Backing up all of my information to floppies was impossible. External hard drives were expensive at the time.

Iomega came out with the external Zip Drive, a small blue box with a fat gray SCSI cable attached. You hooked it up to your computer, popped in a Zip Disk, which was not much bigger than a few floppies stacked together, and a volume appeared on your desktop. Most people think of it as an "icon," but it's called a "volume." You instantly had 100 megabytes of storage. I felt rich with capacity. I could drag tons of my files and music over to the Zip drives, then toss them in a drawer and know that my vital information was backed up, protected, and redundant. A crash would not wipe out hours or days of work. I accumulated many Zip disks

I began to back stuff up regularly. It gave me some comfort. I had spent a lifetime scribbling and still had my notebooks stashed in a drawer. They may have been susceptible to a fire or a flood, but they had endured so far. As long as they were in my possession, I felt secure about their survival. Digital media, on the other hand, could disappear in a moment. When a hard drive dies, everything is gone.

The best plan was to have multiple, redundant copies. Even better to have a few backups and to stash one in another location. I became fanatical about keeping multiple backups of my work. I even printed out all my writing, but soon realized that

I could never print out everything. After taking all precautions, I still worried about things like file compatibility in the future. Would WordPerfect documents for Mac always work? It turns out the answer to that was "no."

The Zip Drive was the first mass storage media I owned. It hearkened back to my cassette days. There was a flexible, magnetic film inside the Zip Disk and you could remove the paper label from the box to write on it. As with cassettes and videotapes, you could label the spine of the case, so you could stack the disks on a shelf and identify them easily. I backed up everything by dragging the files to copy them onto the Zip disk. I could not imagine that most physical media, including Zip Disks, CDs and videotapes, would be dead within ten years. It seemed as though every new storage medium was more stable and had more capacity than the previous one.

I worried about EMP. It's something you know about if you are a conspiracy theorist or, in my case, if you read enough about warfare and technology. It's the concept that a nation could explode a nuclear weapon at low altitude above a city and create an electro-magnetic pulse, or "EMP," which would short out and kill anything with a chip in it, including hard drives and computers. A dark scenario, but still something to contemplate. What could I do about that? A Faraday Cage? I still have not protected myself from EMP.

I worried about hard drive degradation through non-use. I worried about the long-term viability of CDs and DVDs. It seemed as though I was always running in the cartoon while the trail fell away behind me. I still don't feel secure about the long-term viability of my writing and photo files. Will digital media survive? If enough copies exist, maybe.

I began to think about creative heirs, people I could leave my photo, text and sound files to. I thought about the archivists at libraries in the future and hoped they would be familiar with late twentieth-century email applications and word processing formats. Forward compatibility was something I thought about whenever I bought a new piece of software. I became more aware every day of the tenuousness of digital creations.

I learned a hard lesson early on, when I was writing a piece for an online magazine. I created a document one day and worked solidly, taking breaks, then returning to it and trying to make the deadline. This went on for six hours, at which point my computer froze. Freezes happened a lot with the older Mac operating systems. I tried a few keyboard commands and, after a while, had to restart the machine. It was in that sickening moment I realized I had never saved the document. I had only created it. Six hours of work were lost. That was an expensive lesson and it may have been the start of my obsession with keeping backups and preparing for computer disasters. I never made that mistake again.

42

CYBERSPACE

It felt like I was pulled into the ocean. My online life began to take up as much time as my real life. There was still a clear distinction between the two in the 1990s. It became important every morning to wake up and check my email. I would roll out of bed, sign into AOL, listen to the modem noises and the robotic voice saying "Welcome," quickly followed by "You've got mail." I would check my mail and look at the news, but America Online quickly became claustrophobic. I found myself going to the internet more often. Things were happening on the internet and it seemed possible to express myself and make money at the same time.

My job at the bakery began to feel like a desperate waste of time. It had been a good place to rejoin the human race after getting clean, but it was becoming a dead end. I was the wholesale manager and my days consisted of heavy phone work, writing orders with a pen and paper, hounding bakers to complete cakes, boxing hundreds of cookies and pastries a day, and tracking down drivers whose mission was to waste time and avoid detection in Manhattan, Brooklyn and the Bronx.

You could not call them because few people had a cellphone in 1995 or 1996. I knew one person with a mobile phone at the time and he was rich. It was as big as a brick, made by Motorola, and it seemed like a miracle. There was no inkling that everybody would soon own one. People still talked by phone, in person or, increasingly, online with email.

I had dropped out of society for a long time. Now, I began to sew together the broken threads of my past life with my current one. The internet helped me get close again to my brother, who lived just ten minutes from me, and Roland, who was a half hour north. The band had broken up and I had lost touch with everybody from high school or college, let alone the old neighborhood in Yonkers. Aside from my parents, I had broken from my entire past.

An email address made it easy to correspond daily with my brother and Roland. It removed the emotional heaviness of phoning and re-establishing connection. Calling meant going through the protocol of asking how the other was doing and what was going on, circling around the conversation. With email, one of us could just forward a silly link or write "what's up?" and leave it at that.

Roland had spent the previous ten years becoming a master of electronics at his job, working for a company that did custom-built, high-end multimedia installations in the homes of rich people. He had built a serious home recording studio and gotten into computers long before I did. I was impressed. My big achievement had been to not pawn my Gibson Melody Maker guitar while I was at my worst.

At some point, I began to search online for any trace of my college friend, Paul. I found a few possible email addresses and sent out a tentative feeler, along the lines of "If you're the Paul

S. whom I went to college with and did a radio show with…" A few days later, he replied! It was Paul and suddenly a dear friend was back in my life. Years had passed and he was 3000 miles away, but I had found him through the internet. People may take this for granted now, but it felt like a miracle back in 1995.

Paul had gone on to grad school and then moved to California to work in the software and digital video industry. He was on the leading edge of digital video production. Both he and Roland were both doing interesting work with technology while I was driving a pastry truck and managing a wholesale bakery. That did not bother me at first. I was happy to be immersed in life again.

I became restless. When my boss opened another branch nearby and moved us there, in late 1996, I quit. Business was slow and I told him he would be better off without having to pay my salary, which was true. I had just sold a piece of writing for seven hundred dollars and began to dream that I could support myself this way. I ran into a friend, Dave, whose sister was editing an online magazine.

He was getting paid to write articles for them and he told me to send in a query, which I did. I wrote a piece about driving a delivery truck in Manhattan. He was working on a project at AOL. At the time, AOL and independent websites were pouring money into creating content. It was the first inkling that they needed some way to keep people engaged online, and it was similar to what cable TV had discovered – you needed a lot of material to fill up all that time and space.

The amount of content became more important than its quality. Employers began to call us "content-providers," rather than "writers." Dave was working for an AOL content channel. They were throwing different things against the wall to see what would stick. He pitched them on the idea of hitchhiking around the country with a laptop and a cellphone, two futuristic pieces of technology at the time, along with the most cutting-edge device of all, a cellular modem. His job was to hitchhike around the country, post stories, and host chatrooms wherever he landed for the evening.

It was a good idea and he was a good writer, as well as an engaging guy. The only problem was, he was having romantic problems and did not feel like hitchhiking. After he and a friend drove south a few hundred miles and he spent one day on his own, I got a call from him, asking if I would come out on the road to keep him company. We could drive around while he pretended to hitchhike. I jumped at it.

I love the road and everything about it. I like motel rooms, I like waking up in a new place every day, and I love not knowing exactly where I'm going to end up that night. I love the food, the driving and the people you meet along the way. This gig seemed like a good way to get back into writing. I had been writing in my journal on the computer for close to a year, but this would put me with a working writer who was employed by a new media company. At the time, AOL had a cachet that is hard to conjure up now. Equally important, he was willing to pay me a thousand dollars to chauffeur him around for a month.

I had no money saved, but I had re-established my credit and gotten a credit card, so I ran out and bought Apple's best PowerBook laptop at the time, the 5300c. I had two days to prep it before I had to leave. I also needed to pay my bills and find someone to take care of my cat. I spent the next 24 hours installing applications, getting the modem card and dongle to work, and marveling at the experience of working on such a small computer.

The Mac 5300c was thick and chunky. It was made of black plastic and had a low-resolution color screen that looked terrible. It had a trackpad and a "swappable drive bay" that contained a floppy drive. It contained a 100-megahertz Power PC processor and a PC Card slot, which is where the modem card plugged in. At the time, the machine was small. You could carry all 6 pounds of it in a backpack. That model became famous for spontaneously catching on fire, but mine never did self-combust.

I enjoyed setting it up. I liked the satisfaction of installing software and getting it to work, of creating this little, virtual space in which to play and create. I liked fixing things that broke down and I found it satisfying to troubleshoot problems with the software. I installed the copy of Photoshop 2.0 that Louie had given me, even though I still did not know how to use it. The hard drive was 750 megabytes – twice as large as the one on my desktop computer.

The modem and dongle felt tiny. The modem fit into the PC Card slot. The dongle was a small wire with a connector on each end. You pushed the card into the slot, then plugged the dongle into the card. On the other end of the dongle was a phone jack, the standard RJ-11 modular phone jack.

I had to make sure I had the correct settings for the modem, the internet provider, and the computer. When it was all working properly, I would click on a widget in the menu bar, the software would dial, make the connection, and show a little symbol that meant I was connected. I could then check my mail, surf the web, sign on to AOL, and use FTP as I would on a desktop computer. It took me a day and a half to set up that laptop and make it mine. I enjoyed the urgency and the sense of mission that this gig seemed to have.

43

THE FUTURE

When I was done prepping the laptop, I packed a small suitcase, along with my point-and-shoot film camera and a telephone calling card. I said goodbye to my cat and headed over to Hertz Car Rental, on 40th Street in Manhattan, where they rolled out a new Ford Taurus with fourteen miles on the odometer. I got in, consulted my road atlas, and then pointed the car towards the Holland Tunnel and drove.

The previous year, I had taken two weeks off from the bakery job and planned to go out west with the woman I was dating. Instead, we broke up and I flew to San Francisco alone, rented a Mustang convertible, and burned up the asphalt all over the west and the southwest. I came home and told everybody how much I liked to drive. Dave asked me to go with him based on this dubious attribute.

That first day, I drove until Roanoke, Virginia, where I stopped for food at the local Waffle House, then found a motel and sacked out. The next day was a long run that ended 800 miles later, in Mobile, Alabama, during a torrential thunderstorm. I found the Interstate exit motel where my friend was hunkered down, depressed and composing a long email. I greeted him and went to bed.

The next month was an equal mixture of adventure and misery. I had a good time simply because I was on the road and part of an interesting enterprise. Dave was miserable over a girl and homesick, to boot. Each day, we headed to a different town and he pretended to his online followers that he was hitchhiking. Roaming time on the cell modem was expensive, so he rarely used it.

When he did, though, he would have to set up the phone, which I recall as a Motorola that unfolded. He then used a special cable to plug into the modem, which was plugged into the computer. He would go through the Byzantine connection dance of getting the early-generation cellphone to talk to the primitive cell modem and then connect to AOL. It often failed to connect. When it did, though, we had a glimpse, one of the earliest glimpses, of what life would be like for anybody with a smartphone ten years later.

He hosted chats, wrote posts, and answered email. Occasionally, someone in the next booth at a Waffle House would ask what we were doing. When we replied that we were on the internet while sitting there with our waffles, many would shake their heads in doubt. Others did not know what the internet was. This was America in 1996.

Dave also had the earliest practical digital camera – the Apple Quicktake 100, which grabbed my attention. It shot at a resolution of 640x480 pixels, which is terrible now, but miraculous at the time. There was no memory card, so you had to connect it directly to your computer to download the photos. This was another arcane and glitchy procedure. He was supposed to illustrate his posts with photos taken while hitchhiking. He was heartbroken, though, and by the time we reached New Mexico, all he wanted was to go to Memphis to meet his girlfriend for the weekend. In the meantime, despite his misery, or possibly because of it, he attracted a small group of devoted followers and entertained them every night in his chatroom. They followed his travels, enjoyed his columns, and emailed him constantly.

What was I doing this whole time? I was driving a lot. The car was insured in my name, so I became the full-time driver. I became the Robin to his Batman in the chatroom. Under the ruse of being a friend in New York who followed him closely on AOL, I would hang out in the chatroom and bask in his reflected online fame, field questions, and flirt with the women. This devolved into a predictable dance.

The days were full of driving and I was exhausted every night. I had no cellphone and only a film camera, but I had my laptop. It was my first experience with mobile computing. Every night, the first thing I did when we checked in to another seedy motel room was to seek out the phone jack. This was before Wi-Fi existed and before ethernet data ports appeared in hotels or motels. It was before motel owners knew that you might be jumping onto their phone system for anything other than phone calls.

I became an expert in the arcane world of motel telephone systems. The best-case scenario was for a room to have a standard, modular RJ-11 jack and a non-digital phone system. This was not always the case. The older motels had their phones wired into the wall, so we had to carry tools with us to open up the phone jacks. They were mounted on the baseboards, so we had to kneel and connect a modular jack to the wires before we could plug in the modem. When we left the next day, we had to reverse this process and restore the status quo.

Dave was trying to save money, so the two of us stayed in the same room, which meant that only one of us could be online at a time. Since he was the boss and had more legitimate reasons for being online, I deferred to him. This led to me taking long walks along motel strips in small towns. It also led to creative solutions, such as begging motel owners to let me use the phone in an unoccupied room. I would tell them I had to make a call, but I would occasionally tell them the truth, that I was plugging a computer into their phone line. They suspected it would cost them money. If we had not been getting comped by AOL, it would have.

Once we created the physical connection, we had to negotiate the protocol for our modems and that particular phone system. You needed to set up each call differently, depending on what the telephone system required. To reach an outside line and then pause for a dial tone, you went into the dialing string in the modem software, punched in "9" and a comma, then the number. This was common. There were much stranger dialing strings and there were times when we had to trick the phone system by reaching an outside line manually, then quickly plugging in the modem.

Getting online from a motel room in the mid 1990s was a hassle. One night, we stayed at the El Rancho Hotel in Gallup, New Mexico. Their motto, writ large across the front of the building, is "The Charm of Yesterday. The Convenience of Tomorrow." Dave managed to get online just in time for the chatroom and I wanted to join, so I convinced the desk clerk to let me set up in the small switchboard room behind the front desk. I chatted with one of the women from the AOL channel and this became my first romance to cross from the virtual to the real. It did not last long, but it was a great new way to meet someone and to overcome my shyness. Love through technology – the convenience of tomorrow!

I brought along my usual collection of music for this trip and carried it in a small cardboard box. There were twenty-four cassettes, in three rows, stashed under the front seat of the car. Car audio had not changed since the 1970s. I owned few CDs and a cassette player was far more common in cars, especially rental cars. Dave did not like most of my music, so we listened to a lot of local radio. It's less common now, as people have satellite radio, streaming music, MP3s, podcasts and other personalized music options. Back then, though, local radio was ubiquitous and often strange, especially down south.

In Truth or Consequences, New Mexico, we stayed in one of the most hardscrabble, dingy motels I have ever seen. The lobby was filled with broken toys and garbage bags and the walls were bare plywood. There were two outside lines, so only two of the 12 units in this place could dial out at any time. This may have been adequate in the 1940s, when it was built, but it was no good in the mid-1990s.

Four cowboys were in the next room. They pulled in with their horse trailers and pickup trucks in the early evening, went out and bought beer, then drank and watched loud TV late into the night. They woke up at 5:00 a.m. and took off, but one of them was on a long call for hours that night, which left the remaining line for Dave to use on his chat while I walked around the parking lot on the strip in Truth of Consequences.

44

WORKING ONLINE

The trip ended up costing me money, but it opened my eyes to a new way of traveling. I could be online while on the road and maintain a simple version of a blog before the word "blog" existed. I could stay in touch with people through social media and document my travels in real time. "Social media" did not even exist as a common term then, but that's what people were using America Online for.

I did not have all the technology yet - the digital camera or the cellphone - but I had a laptop and I had seen the future. After we got back, all I could think about was hitting the road again, this time on my own. I needed money and I found work doing freelance pieces for web magazines, all of which seemed to hold great promise. At that time, most things on the internet seemed to hold great promise. I was never employed in-house, but these magazines let me hang out and soak up the stardust that even a tiny web startup generated then.

Everybody in the business thought they were on to the next great thing. In a way, they were, even if they had no business experience or literary taste. It didn't matter. If you could start a web site and generate any traffic at all, you were considered hot and exciting. This was the case with the magazines where I was a contributor. For a brief moment in the mid to late 1990s, these sites paid good money for web journalism.

Content became another commodity that companies tried to get for free. They called it "Web 2.0" and web sites would try to get the end-users to generate their own content. It was a great idea from a business perspective, but not so great for the writers. That was still in the future, though. For now, I thought I had found a new career, writing for new media. It was a skill I might have parlayed into a return to the advertising business, but I hated writing copy for the web as much as I had hated writing copy for old media.

Still, I was grateful for the money. It came easily. In the early days, people thought you needed some special skill to write for the web, as opposed to writing copy for tangible media. Somebody would know somebody else and they would refer my name. I would put on a suit, show up at a meeting and get an idea of what they wanted for the web site. We would discuss the sections, then I would go home and grind it out. I was writing copy for the web sites of major corporations before any real sense of style or design existed for the web. Web design was primitive. The use of frames on web pages was the hot new feature back then.

Writing was arranged in sections, which were no more than glorified paragraphs. I would take existing copy and either repurpose it or re-write it. If I didn't know about a subject, I would invent some vague bullshit and write it down, then spruce up the English to make it sound good. Most of my work was accepted with few revisions. I would submit a bill and get paid, having done most of the work from my computer at home. It was the easiest money I ever made, even though the subject matter was ghastly – telecommunications, real estate, camera manuals, and technical documentation. I began to feel almost legitimate. All of this had come about from my purchase of a personal computer.

I thought I could make a lot of money and possibly change the world by writing for the internet. I went to a party hosted by an AOL channel. America Online was putting a lot of money into their "content channels" at this point and trying to remain relevant. At this party, I ended up sitting at one of the computer terminals they set up to show off their new channels.

At that time, you could spot internet junkies by the way they would gravitate to any live connection, jumping into chat-rooms and sending instant messages. People were not walking around with smartphones stuck to their faces, stuck in a cyber vortex all the time, simply because the phones did not exist. If a computer happened to be around and it was networked, you grabbed it. Getting lost online in public was almost cool then, rather than a source of minor shame.

I found myself alienated, as I usually am at these events, and sitting at a computer next to a tall, beautiful woman. She introduced herself and I awkwardly reciprocated. Screen names were exchanged and, within five minutes, much to my amazement, we were flirting online while sitting next to each other in the middle of this public event. It was exhilarating and it was possible for me only at that moment in time.

Soon enough, we started dating. She was an actual model, a beauty pageant winner, and quick-witted. She was also half a foot taller than I. The whole thing was strange. She hired me to create my own site on AOL, which I did. I tried to migrate the population from my Dave's defunct site over to mine. It failed. I had no hook to offer, aside from my introverted personality. The online world was changing quickly. No longer could you get paid simply for existing online. You needed to offer something. This woman and I broke up and she eventually laid me off. That was okay, I was ready to move on, having apparently been broken of the need for love and cohabitation a few years earlier.

45

CELLPHONES

I GOT MY FIRST CELL PHONE IN 1996. THE NOKIA 2160 was the size of a fat candy bar, dark gray, with a standard keypad on the front, a few navigation buttons, and a small yellow screen with gray characters. A stubby antenna stuck out of the upper right side. There was a removable battery on the back. The day I bought it, I called some friends and my parents, just so I could say "I'm talking to you from my cellphone." It could save 100 phone numbers and it had call waiting. It was one more wall wart AC charger to find a plug for in my house and to carry while traveling.

I loved to take road trips and would fly out to California or Las Vegas, rent a car and then drive. I had always done this any time I had gotten two weeks off from working at the bakery. I lived for those two weeks. I had read Thoreau, especially his admonition that one should ideally work for a month of the year and then take the next 11 months to wander. I had spent enough of my life in cubicles to know that I wanted to wander more. I had witnessed, with the America Online fake hitchhiking trip, how you could harness a cellphone to bring the internet to your laptop. I saw how you could take your online life and carry it with you on the road. You could then take the road and put it online.

I got that first cellphone as a safety measure on these road trips. It was a dumb idea. There was almost no cell service outside of the big cities then, so it would have been useless if I had broken down somewhere. My carrier was AT&T Wireless and that phone carried an outrageous plan – 30 minutes of talk time for a monthly fee, after which you paid by the minute. By the mid-1990s, cellphones were more common, but they were not something you saw every day. At that time, when you got a cellphone, you would tell people "I have a cell number now." It was a marker. For some, it may have been a status symbol.

At first, it was convenient to be reachable at all times. My ultimate goal was to use the phone as a modem for my laptop, but that would take a year to make happen. It was the beginning of the cellular explosion. I carried the little Nokia phone on my next trip, made and received a few calls, and that was it. I was too worried about racking up expensive minutes to enjoy the convenience, but I did enjoy the novelty.

At home in New York, I hung out with my friends at the ALT. (pronounced "alt-dot") coffee shop on Avenue A most nights. Not much happened there, but I liked to watch everybody sitting around online at computer terminals that they paid for in tiny increments. The bathroom was legendary, not so much as the bathrooms at CBGB's, but still well-known. It was filthy, with graffiti all over the walls in the old Lower East Side style, and a toilet that usually worked. Most notable was the old bathtub, which was filled to the ceiling with broken computers, mainly tan boxes and monitors. It was a cross between an art installation and a general statement about malaise and technology.

In 1997, Bill Gates and Steve Jobs made peace. Bill invested in Apple and agreed to keep making a version of Microsoft Word for the Macintosh. It was a critical gesture and it ensured that Apple would survive. It also took some of the heat off of Microsoft and accusations that it was a monopoly. More and more of my friends owned computers and were going online. The internet was changing from a diversion to a necessity. Most people were still paying their bills with checks and money orders at this time. Classified ads in the newspapers were still the norm, as were personal ads. Online shopping was becoming a thing, but many people would not yet dream of ordering household items from a website.

I had a computer, so there was no lack of people calling me and wanting help with their resumes and other things. My friend Roger, who worked for a shipping company by day and went to sex clubs by night, heard about the internet and begged me to post a personal ad for him on an S&M dating site. I obliged him, but then found myself in the middle of an exchange between him and a woman who answered his ad. She would send me an email and then I would have to call him for a response, the whole time telling her I was doing this for him. It was an odd situation and one that I had to get myself out of, but not before it grew crude, sordid and sick.

I took most of it and turned it into a story for some online magazine. Everything in my life was grist for the mill and I was writing for writing's sake. A literary agent saw one of my first online stories and emailed the web magazine with a request that I contact him. The publishing world must have seen the internet as the genesis of a new literary farm system, something that just landed from the skies and promised a crowd of new talent.

I could not believe an agent was interested in my work. He wanted to know if I could write a book based on my experiences. It would be another New York drug memoir. I was interested, but the pressure of having to come up with a proposal, combined with him vetting it, making edits and suggestions, and having to hew to a deadline killed it for me. After writing a few dozen pages and procrastinating for a year, the project died.

46

TEACHING

BY LATE 1997, I FOUND SOME TEACHING WORK. A FRIEND gave my name to the acting dean of the Communications Department at a college in Long Island. They needed someone to teach a course called "Writing for the Internet." I jumped at the chance, even though it meant a two-hour drive each way every week. The class was a success and I learned how to teach while on the job. The next semester, they gave me an additional course, called "Multimedia Criticism," which also went well. This work led to adjunct positions at other schools, such as Long Island University, in Brooklyn, and the New School, in New York City. At the New School, in 1999 and 2000, I taught a course called "Internet Journalism." The description in their course catalog read as follows:

Internet Journalism: Writing for the Web

A computer, a modem, and a phone line can change your life. This course, for beginning and advanced writers, teaches everything you need to launch an online career and pitch stories to online magazines. Many writers are finding opportunities in this new and burgeoning field. The class explores Web publishing sites, chat rooms, creative home pages, and boards, as students submit work for publication on the Web and write for their own Web sites. Learn the particulars of writing for online media. Guests may include online publishers and editors.

It was a completely unwired classroom, which made it a challenge to teach about the internet. I lectured the students on what I knew, then we would talk about issues related to writing for the web, after which we talked about their writing.

The lesson plan from that class looks dated now, but it was appropriate for the time. One of the things on my list to discuss in the first class, after the usual round of introductory comments, was "What's a URL?" I still wonder if people know what a URL is. Then again, does anybody need to know that acronym if they don't create web content? I told them to carry a notebook and pen everywhere.

The administration finally wired up some classrooms, so I got what they called a "smart classroom" with a projector in it the following year. There was a hunger to learn about the internet. People began to realize that it was a big deal. When I hung out at the offices for one of the web magazines I wrote for, people would email or call to ask if they could come by, just to be around something dynamic and new. I remember a middle-aged visitor who called to say, "Can I just sweep the floors there?"

In the 1990s, people still lived most of their lives offline. We were not connected all the time unless we worked in technology or some place with a specific need to be online. Early on, though, people began to talk about "internet addiction." Most people didn't care. We did not sit around in class and question the subtleties of separating one's life from one's presence online. My job was to teach something that may not even need to be taught anymore.

I began to teach classes on how to use Photoshop, Dreamweaver and After Effects. The New School course evolved into something else and I never got tenure. My days in the adjunct gulag had begun, but I was happy. Being online seemed to add an extra dimension to life. It was not woven into daily experience so much as an escape from daily experience.

By now, the number of pornography sites had begun to explode. I might be sitting around, having a bite with friends, and we would all compare notes on what we had discovered online. At the time, a site called "askjolene.com" was big, as was "Persian Kitty," a site filled with porn web links. I was not into porn, but it seemed as though everybody had a favorite site or two. Porn sites, initially, at least, were the only sites able to turn a profit on the internet. They began charging for access. When competition heated up and advertising became profitable, they became free, but they led the way in the paid-subscription model.

Soon, my friends were telling me about specialty sites they had found, from hardcore dating to specific fetishes. One of my writing gigs was interviewing people about their jobs. One of my subjects called herself an "Adult Webmistress." At the time, it was an exotic job title and she fashioned herself after a brothel owner out west in the late nineteenth century. At first, the internet allowed people to reinvent themselves.

By 1999, online commerce companies began to compile dossiers and databases on individuals, collecting personal information and buying habits. That year, the CEO of Sun Microsystems told a product show crowd "You already have zero privacy — get over it." By the year 2000, after it came out that the DoubleClick internet advertising firm had collected private information on 90 million households, there was talk about instituting privacy standards. Not only did that not happen, but things got much worse.

I was using the Altavista search engine, which had a clean, simple interface. It seemed to bring back search results in a smarter way than previous search engines. The web had changed from a quirky place to find cool stuff to an ocean of information. I would think to myself, "somebody is going to figure out how to make money off of this and it'll ruin everything." It had already happened, and it happened so fast that I was blind to it.

47

THE GRAND WAGONEER

I MADE ENOUGH MONEY THROUGH FREELANCE WRITING to buy a car. I had been walking by an old gas station and mechanic shop on the Bowery for months, partly because it was on my way home from the Lower East Side, where I ate every night, and partly because of a video game. An arcade version of "Gravitar," a game I had known and loved in the 1970s, appeared one night outside of this garage, like a dinosaur bone washing up on the beach.

I was shocked to see it the first time, this artifact from the early days of the video game revolution. I began to play a few rounds every day on my way home. This was a simple, vector-based game that Roland had been especially good at thirty years earlier. It brought on severe tech nostalgia. One day, while stopping by to play the old arcade game, I noticed a beat-up Jeep Wagoneer parked next to it with a "for sale" sign in the window. I was looking for just such a vehicle to take on trips out west.

I was almost forty and had never owned my own car. I had been trolling the new 4X4 websites and reading magazines about off-roading. You could still find a lot of specialty magazines at the newsstands. I had been thinking of a Ford Bronco II, but it did not feel right. I had also been thinking of a Jeep Wagoneer and suddenly began to see a lot of them on the streets of New York City. For years, I had been flying out west and renting a car, but that was a limited way to travel. I wanted to explore and to go off road. I wanted my own car.

The Wagoneer was the last American vehicle produced with a carburetor, as opposed to fuel injection. Jeep had designed it in the early 1960s. It was old technology. The heavy leaf spring suspension was primitive. Jeep had been sold to American Motors Corporation, who made my Wagoneer, and they added some of their own parts in order to save money. Everything about the basic design was old, but it had just enough new technology to cause trouble.

The electric door locks were from the first generation of automotive electric door locks. The joystick switch that controlled the power mirrors was also from an earlier mechanical generation. When it broke, I searched for months to find a replacement, and even that was a used module that I found on eBay. I had the same experience with the electronic key switch on the rear window. The basic vehicle was sound, though. The low range gearing was activated by yanking on a lever that was mechanically coupled to the transfer case. No electronics were involved. It was old technology at its finest.

I was friends with a woman named Peggy, who was an art student and would cat-sit for me when I went away. She and I were both obsessed with the road and took a series of long road trips out west, places where there was no electricity or phone service, let alone internet service. I was getting back into photography at this point and was drawn to landscape work. I liked to shoot old buildings, abandoned stuff, detritus, signs and motels. I especially loved old motel signs. I thought I was the only person doing this at the time. Then I discovered Walker Evans. A while later, I discovered Stephen Shore and Saul Leiter, who had both photographed similar subjects.

I was shooting film and had to use a tripod for night shots. We would leave New York and wander across the country, stopping and shooting everywhere. When we returned to the city, we sent our film out to be developed. You had to specify matte or gloss, borders or no borders, for the prints. We preferred gloss and no borders. Then we would wait for the photos to come back. When they did come back, Peggy and I would meet at the Kiev Ukrainian restaurant, in the East Village, and go over what we had shot. We would usually order double prints so that we could each have a set. Getting them back and looking at the prints was half the fun.

I joined an online forum devoted to the Jeep Wagoneer. By now, I had also joined a forum dedicated to Fender Telecaster guitars, where I bonded with other guitarists over the Telecaster, a mid-twentieth-century technological marvel made from a piece of maple, a slab of ash wood, and some electronics. I joined another online group, devoted to digital photography, which was in its infancy. You could join these groups, which were text-only, by email. Once approved, you could post and read other posts through digests that you got as email. Some called them a "BBS," for "bulletin board system," and some people called them "electronic bulletin boards." They were a holdover from the early days of the internet.

By 1998, I was been using the Hotbot search engine. I have always been brand loyal, but it was difficult to gin up a fondness for search engines. The tech websites would review the latest one and extoll its benefits. I probably read an article about Hotbot and began to favor that over Altavista. When the buzz began to build around Google, later that year, I checked it out. Their whole premise was to rank pages according to how many other significant websites they were linked to. Google delivered better search results.

I liked the interface and their "Don't be evil" motto appealed to me. This was before they became the face of Big Data. This was before some writer, somewhere, had the revelation that the users of Google were the actual product. At the time, the idea of indexing all of the world's information seemed noble.

In my AOL days, I had become acquainted with the psychology of online social groups. My experience in these new groups confirmed what I already knew – people act worse online than they do in the real world. These groups all had forums where members could post things and engage in debate. The debate would inevitably descend into name-calling, abuse and trolling.

This happened with forums related to guitars, cars, computers and anything else that people discussed. The anonymity of online personas brought out the worst in humanity. It fostered hostility, rudeness and incivility. People would say anything from behind the veil of a screen name. Unlike groups that meet in person, an online forum tends to have members from all different social circumstances, many with opposing viewpoints.

Time and distance were shattered online. The internet was a great equalizer. I found that people who loved Jeeps or Telecasters were just as likely to be conservatives as they were to be liberals. They might be black, white, rich, poor, male, female or transgender. Most of the people in these groups were male, though. You did not always know who you were talking to. This was when the meme started to circulate: "on the internet, nobody knows you're a dog."

I used the Apple 5300c on all those road trips. That machine carried me for three years of hard use, after which I gave it to Peggy. She used it for a few more years. By 1999, I was one of many people eager for Apple's next model. They obliged us with the PowerBook G3, a more powerful machine in a sexier form factor. It had sleek lines that made it look like the hood of an old Pontiac. The Mac blogs were lighting up for months with speculation, arguments and trolling. Somebody on a forum would proclaim his undying allegiance to the beautiful "Blackbird" PowerBook, the 500 series, which predated my 5300c. They would rave about this machine, the first with a trackpad and also a lovely piece of industrial design. On internet forums, an emotional attachment to technology can override fact.

I got the next version, codenamed "Wall Street." Hardware geeks and commenters like to refer to new machines and technologies by the secret code names that companies use with their projects. This laptop was much more powerful than my PowerBook 5300c. It had a swappable bay on the left side, so I could pop in an extra battery, a floppy drive unit, an Iomega Zip drive or even an extra hard drive. That was cool and I took advantage of it. I once again experienced the joy of setting up a new machine.

I replaced my original cellphone, the Nokia, with another model, a Motorola StarTAC, and that was when things began to get interesting. The StarTAC was a better phone and it looked more futuristic. It was smaller. My first phone carrier had been AT&T, but I switched to Sprint when I bought the

StarTAC. They were offering better deals – more talk minutes for less money. You could not port numbers, so I had to get a new number. This was no great loss, as most people still considered your landline to be your main number. You gave your cell number to people you cared about and used your home phone for business and identification purposes.

The Sprint network was terrible, especially outside the city. My second StarTAC phone was silver and it was dual band. I purchased an extended-life battery, which made it less svelte. With this phone, I moved to Verizon, whose network was better. My phone number changed again. I had to wait for my contract with Sprint to run out, as the carriers charged huge fees at the time to leave them. I was starting to rely on phone access at all times. The change from a landline to a cellphone was imperceptible at first.

This was when I was first able to send SMS messages – the acronym for "Short Message Service." At the time, you could send an SMS, soon to be known as a "text message," only to someone on your own carrier. I knew few people on that network and fewer who would want to text. You had to type your message on the numeric keypad, so typing individual letters took multiple clicks. That was a hassle.

Texting was a novelty at first. When they finally fixed the technical issues and you could send texts across carriers, the number of people texting skyrocketed. Texting had a special appeal to people who did not want to deal with voice calls, people who only wanted to send a quick message, and people who could not talk because they were at work. For those and other reasons, texting exploded into a new form of communication, especially after the year 2000.

The same people who preferred texting were the ones who loved Caller ID, a service you could not get until the late 1980s and which cost extra on landlines. On most cellphones, it was available as a free feature. Now, anyone could screen phone calls. As a young child in the 1960s, I had watched my mother come up with a workaround for screening phone calls. If she did not want to talk to somebody, some of her friends would know to call our house, let it ring once, then hang up. If they called back immediately, she would pick up. She called this "the signal." My mom did not live long enough to get a cell phone, but I'm sure she would have loved them.

I bought a connecting cable for the StarTAC. It let me plug the phone into a PC card modem. This allowed my cellphone to connect to the internet over a cellular network. It required arcane settings on the computer. You had to program the dialing software precisely or it would not work. Calls dropped often. The connections were slow. When it worked, though, it was magic.

I would pull out my cellphone and laptop by the side of a desert road or in an airport, where Wi-Fi did not exist in the late 1990s, set it up, dial, and then check my email. If I had time, I would check out news on the web. It was like having access to your own private world while everyone around you was stuck in reality. The novelty wore off over the years, as connectivity became ubiquitous. In the early days, though, it felt special, this coupling of cellphone technology and computing. Cellphones then were strictly for voice calls and minor texting. "Smartphones" were a mirage in the future. Dialup connections on the road were so slow that I had to routinely call friends and tell them "please don't send me attachments while I'm on the road."

I bought my next desktop, an iMac, in 1998. Apple had been a marginal company for years and I worried about owning a machine that would no longer be supported after they went bankrupt. When Steve Jobs returned, though, Mac users had hope again. They introduced the first iMac, in Bondi Blue, and people celebrated with a midnight parade of brand-new VW Bugs in San Francisco.

It was a more innocent time. I bought one. It was a much better machine than the old Performa, in a much smaller package, and much cheaper. This was confirmation of Moore's Law. Moore was the Intel co-founder who said that computing power doubles every two years. That's an oversimplification and it's not a physical law at all. Some people go to great lengths to deny or debunk it, but it was a useful benchmark and it confirmed what I had seen thus far in my life – that every piece of technology got faster and smaller with each passing year.

In the early 1980s, you could buy external hard drives for your computer. They were as big as a small filing cabinet, made a lot of noise, and offered you a few dozen megabytes of storage for a few thousand dollars. As the raw processing power of CPUs increased every year, the simultaneous great advance was storage that got cheaper and smaller in huge leaps.

The iMac was an all-in-one unit – the monitor, computer, and modem were combined. It looked unlike any other computer that came before it. It was a large, shiny, brightly colored bubble with a handle on top. It was the first major design from Apple's designer, Jony Ive. It had a built-in 56.6 kilobyte-per-second modem. This was much faster than the 14.4 kilobyte-per-second modem that shipped with my first Mac. It was the first Mac to feature USB ports and get rid of the floppy drive.

The previous few years, there had been a speed race among modem makers. Dial-up internet would soon be eclipsed, but modem speed was still an important computer specification in 1998. The "i" in "iMac" stood for "internet." It had a fast CD-ROM optical drive, the pixel density of the display was denser, and the processor was faster.

The USB ports were smaller and simpler than SCSI ports, as were the USB cables. It made my life and my work better. The big flaw in this machine was the infamous "hockey puck" mouse, a round mouse which I never got used to and which some point to as a red flag. They argue that Apple put form over function with this mouse. Maybe they did, but it was not the first time. Steve Jobs had been obsessed with form and design since the beginning. I bought the iMac with money earned from working on the internet.

I had a few friends who were getting into personal digital assistants at the time. Each one of these devices had diehard fans, but nobody surpassed the intensity of Blackberry users. The Blackberry devices, from RIM technology, a Canadian company, felt good in the hand. The first device to be called a "Blackberry" was the Model 850, an email pager, released in 1999. Most Blackberry devices had a physical keyboard and a scroll wheel on the side. After 2001, they functioned as telephones, calendars, and organizers, but their success was due to an early focus on email.

The Blackberry was the first device to nail sending and receiving mobile email. It was revolutionary at the time. Nobody knew where you were when you were emailing them. Some Blackberry users had rapid-fire email conversations that resembled texting before texting even began. Blackberry devices did not render documents accurately, though. They were not WYSIWYG, pronounced "wiss-ee-wig," the acronym for "What you see is what you get." They operated securely on the Blackberry network and were encrypted from end to end. This was a big selling point for enterprise - businesses and professionals. I was content to not have constant email. I already felt as though I was hooked in enough to the internet.

48

PHOTOS, PHOTONS AND PIXELS

I WAS STILL USING FILM CAMERAS FOR SERIOUS SHOOTING, as were most other photographers. In the early 1990s, I bought a point-and-shoot Olympus film camera that quickly broke and was not worth getting repaired. I replaced it with another point-and-shoot, a little Yashica T4. I was not thrilled with the picture-taking experience, though. I wanted more control and different focal lengths instead of a fixed lens.

In 1997, I wanted an SLR again, so I went to a used camera store on Christopher Street in Greenwich Village. It was the kind of place that would not exist in another ten years, after rents went through the roof. A broken-down middle-aged guy, with wire frame eyeglasses straight out of the 1970s, owned it. He sold me a dented, old Olympus OM-2n, a more sophisticated version of the camera I had pawned in the 1980s. I bought three secondhand lenses and started to shoot more photos.

This rig carried me through a year or two of shooting, until I foolishly set it down one day on the hood of my Jeep Wagoneer. The motor was running rough on a freezing day near the town of Deposit, New York and that camera vibrated off the hood and hit the ground hard. The following week, I bought a used Nikon F3, which began my descent into the world of Nikon lenses. My next few trips saw me carrying that F3 and many rolls of film.

I began to play music with my friend Louie, the guy who had given me all the shareware for my first computer. I pulled my Gibson Melody Maker out of the closet, put on new strings after ten years of disuse, and brought my solid-state Peavey amp from the 1970s to his apartment. Louie had a few friends over for a regular jam and it gelled into a reggae band, fronted by a woman from Iowa and a ragamuffin rapper from Barbados.

I played skank and shadow guitar. Skank is the clickety-click rhythm guitar in reggae and shadow is a muted bass string picking on the guitar. I was obsessed with echo units and tried many of them. I also moved from the Gibson to a Fender telecaster, whose bridge pickup gave me the sound that I needed. We ultimately traveled to California and then to Iowa to play a bunch of gigs and a festival.

In addition to the guitars, amps and effects, I traveled with a flip phone, a laptop and my first digital camera, an Olympus D200L. That Olympus cost a lot of money for me at the time, close to eight hundred dollars, which was about a third of what I earned all semester teaching one class. The files it produced were anemic – 640x480 pixels. It took no memory cards. The internal memory could store only 20 photos at a time. It used a serial cable connector, so getting it to communicate with a Macintosh computer took some fidgeting and a special adapter. Still, it was a digital camera and that made it special.

Film and digital cameras both capture light, carried by photons, to create an image. Film varies according to the chemicals that coat it. It can be color or black and white. It can be "fast," meaning it captures light with a shorter exposure, or it can be "slow." Film can also contain many different varieties of graininess that are properties of the coating chemistry. When light hits the film, this creates the image.

Digital cameras use a sensor to capture the light. You expose the sensor through the lens, just like film, and it converts the light into digital information. Digital camera sensors are measured in "megapixels." A "pixel" comes from the words "pix," for "pictures," and "element." A pixel is a picture element. One digital image on a monitor is made of many pixels. A megapixel contains a million pixels. That first Olympus camera, with its tiny sensor and tiny photo files, was far less than one megapixel. Still, it was magic.

I could shoot a photo and look at it immediately in the small screen on the back of the camera. After shooting the 20-photo limit, I would hook the camera up to my computer and set the capture process in motion. I had to import, name and save each photo. After that, I had to erase the camera before I could take the next batch of photos. Digital camera storage was being treated the same as computer storage.

It would take a few years before the quality and resolution of a digital camera could approach that of film. Some people argued that film would always be superior. They blogged and ranted about it on internet forums and in user groups. Many of us saw this camera and the Apple QuickTake for what they were, though – the first shots of a revolution in photography. I would download photos to my laptop, dial up the internet with my cellphone modem, and send them to people.

Facebook did not exist, so I shared photos via email. This was cutting-edge at the turn of the century. A film camera had better resolution, but you had to get the film developed before you could see it and then you had to scan it before you could use it digitally. The ability to shoot and quickly send a digital image was revolutionary.

Photography had gone through one technological innovation after another, but had been stuck with chemical-based emulsions for a century. When my mother died later that year, in 2000, I held on to her box of negatives. Most had been taken with the Instamatic cameras that she favored for 20 years. I would later grab the metal boxes that contained my father's slides. These were precious, irreplaceable artifacts from my family, my youth, and an earlier era in photography.

Digital cameras were small computers with lenses. You would capture the light with a sensor, then convert it to binary code with the circuitry in the camera. By then, people knew that cars were filled with microchips. We also knew that some computer chips contained many microchips and that these were called "microprocessors." Marketing people told us that the main chip in a computer had so many circuits that it was really a "computer on a chip."

For the next few years, the digital camera world was like an arms race. New and more powerful models came out all the time. A world of websites blossomed into existence. These predicted coming models, dealt in rumors, and reviewed new hardware. In 1999, there was a lot of buzz around Nikon's Coolpix 950 camera, a brilliant design that resembled two small boxes, linked together and pivoting in the middle. I bought one the day it was released.

The 950 behaved like a real camera and it took great photos. It had an elegant design and was built with a magnesium alloy body, which felt solid in your hand. It held a removeable CompactFlash card, so you could take unlimited photos before having to touch your computer. The memory cards then had much less capacity than they do now, but they were building bigger and faster cards all the time.

The cameras from one generation might not work with the cards from the next generation. An older camera might not work with a newer card. You needed more powerful computers to handle the bigger files from the new cameras. Software would improve, but it would not run on old computers. This cycle created a seemingly endless need to upgrade one's gear.

I carried the Coolpix camera everywhere. It was the first practical digital camera that I owned. The photos were great, but they still could not replace the resolution or dynamic range of film. When the Coolpix 990 came out just a year after the 950, I again bought it immediately. The Coolpix 990 was a better camera than the 950, although similar in size and form. As with the 950, you could twist it in the middle and look straight down into the viewfinder, like an old Rolleiflex. People argued about it online. The real buzz started over the next big thing – a digital version of a single lens reflex camera, or a DSLR, a "Digital Single Lens Reflex."

In late 1999, less than two years after I had bought the first iMac, I upgraded to an iMac DV SE, for "Special Edition." It was translucent gray and white, had a more powerful CPU, more RAM and a slot-loading optical drive, meaning the CD or DVD went into the slot like a toaster. No metal housing came out to receive it. It had a faster system bus and video card. More importantly, it had Firewire 400 ports. This was the newest and fastest way to connect peripherals to your computer. I gave the Performa to somebody who needed it and sold my first-generation iMac.

Friends knew of my Mac tech prowess and would call on me to put out technical fires. I never charged people money to help them. I was an evangelist and was thrilled just to help them solve problems with their Macs. Eventually, some would give me their old machines as a form of payment. I would wipe these machines clean, restore them, and get them into the hands of somebody who needed a computer. It was an arrangement that made everybody happy. I would always tell people to back up their computers. I didn't want to sound preachy, but I had lived through data disasters. They would look at me like I was crazy. If I was crazy, I didn't care, just so long as my data was safe.

49

Y2K

THERE WAS A GROWING FETISH WITH INTERNET connection speed among tech geeks and regular citizens. A professor friend of mine told me about the speed of the T1 connection at her school and I was envious. I was still using a dialup connection at home and not spending any personal time on the computers at schools where I taught, although I was teaching more and more. Nobody I knew was carrying a laptop around and using Wi-Fi in the late 1990s. If you saw people in public on computers, they were at an internet café on a desktop computer.

I was still teaching at the college in Long Island, driving there on the Long Island Expressway every week in my Jeep Wagoneer, playing cassette tapes over and over on what the somewhat sleazy car seller had told me was a "thousand-dollar sound system." I picked up a class or two in Brooklyn, as well as at the New School, in Manhattan. In Brooklyn, I was teaching Adobe Photoshop and After Effects. Adobe Dreamweaver was my least favorite subject. The mistakes that the students made in Dreamweaver were the worst for me for me to fix and explain. I avoided teaching web design classes after that, especially since the technology was changing fast. One year's lessons were usually outdated the next year.

In all of these classes, I relied on a computer hooked up to a projector. I sat in the back of the class and took students through the lessons on the projected computer. The students sat in rows in front of me, each at an individual workstation. We were on Windows machines and I had to get used to an unfamiliar user interface. Some of the students were computer illiterates and I had to show them the most basic concepts of creating and saving documents, as well as copying and pasting, and more. I presented lessons, tried to make them interesting, answered questions, and gave them assignments. Then, I would flick the projector off and surf the web while they worked.

Most of the classrooms were low-tech. I tried to make sure that students who were unfamiliar with technology felt comfortable. A few years later, in my New School class called "Politics, Religion and Media," we would discuss the popular term of the day, the "digital divide," that separated the well-off, with their devices and their ability to pay for access to the internet, from the poor, who were struggling simply to survive. Many of my students had no computers at home.

The New School was still catching up with technology when I started there. Even after they began to upgrade their systems, I still had to request a "smart classroom" before every semester or I would end up with a room that had no internet connection or projector. The regular classrooms had no computers, just a big table surrounded by chairs. If I were not in a "smart classroom," I would request a projector ahead of time and some diligent information technology person would wheel in a cart with a projector on it, hook it up, test it, and then leave me. It was a holdover from the days of the "audiovisual presentation."

Even when it all functioned, the connection would be unbearably slow. Web pages would not load. After a semester of technical problems with one class, I brought in printouts of web pages, just to make sure we had something to discuss in case the machines didn't work. Students had no smartphones to pull out in class and nobody took notes on a laptop, even if they owned one. They used paper and pen, as I did at the end of the twentieth century and the start of the twenty-first century, which everybody was starting to call "Y2K," for "Year Two Thousand."

There was a certain amount of hysteria associated with Y2K. Aside from a natural appeal to the doomsday crowd, the apocalypse fetishists, the religious extremists and the history buffs, Y2K was of real interest to the IT people and geeks of the world. "IT" is short for "information technology" and those who work in it are the people necessary to keep the brave new world up and running.

In earlier days, well into the 1970s and 1980s, it was less complicated. The average office would get the phone company technicians in repair the phones. They would need to send their typewriters out to the shop for repairs. By the second half of the twentieth century, they might need visits from the photocopier technician or a service tech from Pitney Bowes, the company that built postage meters, but that was it.

Offices and businesses could operate with basic knowledge and a few common repair skills. Stores had cash registers and machines to take credit card impressions, but nothing was integrated or wired together. None of these machines spoke to each other in any way. There were some exceptions in the fields of advanced research or finance, but most places where people learned and worked were untethered from any greater network. The only network that most people used was the telephone network.

By the year 2000, most offices, businesses and classrooms needed full-time professionals to maintain their computer technology and related communication networks. Y2K threatened this infrastructure with some serious potential hazards. The public was bombarded for months, with predictions ranging from minor to catastrophic. It was said that digital clocks designed in the twentieth century would stop working accurately, that spreadsheets used in the financial markets would cease to perform, that machines would not know what time or date it was. Systems that governed electricity, water, hospitals, automotive components – everything was predicted to fall apart and leave us in a dystopian state. People were advised to get money from ATMs ahead of time and to stockpile food and water. Planes might literally fall out of the sky. Anything could happen.

People took these threats seriously to greater and lesser degrees, depending on how much of the survivalist instinct they had in them. I'm sure that many people became anxious, whether they prepared for Y2K or not. Combine that anxiety with normal year-end expectations and it became the main topic everywhere during the final weeks leading into the new year. I was playing with the reggae band. We had our biggest gig to date on New Year's Eve of Y2K, earning a few thousand dollars to play all night at a club in Greenwich Village. I had a plane ticket to fly out west early the next morning and meet my friend Peggy in Phoenix.

We were planning to travel around the west, taking photos, soaking in hot springs and living in an RV for 10 days. I thought, if I had to go, crashing in a plane on Y2K was not the worst way. We played the gig, three or four long sets, and I made it home with my guitar and amp in the middle of the New Years' insanity. I said goodbye to my cat and headed to the airport and greater glory. Nothing happened, but people took notice, however briefly, that technology systems could have a dramatic influence on their lives. The coming century would drive this home again and again.

50

"INTERACTIVE TELECOMMUNICATIONS"

EVEN THOUGH I WAS TEACHING AT A FEW DIFFERENT schools in the city by this point, I could never get out of the adjunct gulag. I would teach a class for a year or two, but never get a permanent position. I was often at the mercy of administrators or department heads who gave the jobs to their friends and family. I made a living teaching and working freelance writing jobs. I was on the computer all day, reading and writing.

My life revolved around working, trying to take as many road trips as possible, and watching baseball on cable TV. I walked around with a small transistor radio, listening to games when I was outside. I played with the reggae band a few times a week. I embraced older vacuum-tube guitar amps, mainly those built by Fender. I began to think more about older technology. Between my infatuation with cameras, guitars and computers, I realized that I could be labeled a "gearhead." It may have sounded like a derogatory term, but I didn't care. Part of being an artist is knowing and maintaining your tools.

I had a few girlfriends after my marriage ended, but nothing that lasted for long. Online dating services were new at the time and growing fast. I tried them out, but quickly became disillusioned. I had my cat, my music and my writing. After my mom died, in 2000, I decided to go to graduate school and get a degree. It seemed like the right thing to do. I was 40 years old. A growing sense of mortality and the hell of adjunct life propelled me into getting an advanced degree.

I heard about a program at NYU called the Interactive Telecommunications Program, or "ITP." Even though the idea of "telecommunications" seemed to reflect the forward thinking of an earlier time, the program had a reputation for doing interesting work and covering subjects that I wanted to learn. The school was close to where I lived and offered a degree that was respected. A friend told me "they're an assembly line for new Microsoft hires," which was not at all what I wanted and, it turns out, was not accurate, but it seemed like a good endorsement at the time.

I applied and got into the program and was excited to return to school. I was hoping to meet some like-minded souls and learn something new. Over the next two years, I would learn about robotics, get an introduction to writing code, and study user interface and information design. I learned about non-linear video editing and digital audio. Most importantly, I learned that I could teach myself new things.

51

9/11

My second day of class at ITP was September 11, 2001. As I walked to school that morning along Washington Place, I stopped at the southwest corner of Sixth Avenue for a traffic light. At that moment, a fire truck came roaring down Sixth Avenue in the wrong direction, something you don't see often on a busy avenue at rush hour. All the firefighters were standing in the truck with their bodies jammed out the windows. They were all staring downtown.

My head turned with the truck and my eyes followed theirs until I saw the twin towers of the World Trade Center. The north tower had a tremendous gash high up and black smoke was pouring from it. I pulled out my Nikon F3, which was loaded with fast black and white film, and snapped a photo, then I walked on. I was fixated on going to school and doing well academically. This seemed like a terrible situation, but it was being addressed. Firefighters were on the way.

I went through Washington Square Park and stopped at the fountain to take another photo. Small groups of people stood in silence, watching the tower smoke. A homeless person was screaming about the end of the world and divine retribution. I moved on and took the elevator up to class. My fellow graduate students were standing around and talking about the fire in the tower, but we all went into class. The subject was graphic design. The teacher had us make the obligatory introductions, so we went around the room and said something about ourselves.

A fellow student who had been a reporter for Reuters kept leaving and returning, checking the TV news feed in the common area. We all sat around the table in class, vaguely troubled, but going through the motions of a first class for the semester. As the teacher was describing what the class would involve, the ex-reporter burst into the room and screamed "The tower just came down." The teacher slowly said, "I'm not sure what to do," as people slowly stood up and left without saying a word. I thought "Welcome to the twentieth-first century" and walked out into the common area with everybody else. Nobody knew what was going on, but a terrible feeling was hitting us all.

I took the elevator downstairs and saw a river of people walking north in the middle of Broadway. Some were covered gray and white soot. I wandered over to Saint Mark's Books and bought the etymological dictionary I had been looking at for weeks. I felt like an automaton. I was the only person in the store and the lone clerk seemed like she was also in a dream. Outside, people were clustered around a delivery truck, listening to the news blasting from the truck's radio.

There were no smartphones. If you were in the street, the only way to get the news immediately was communally, gathered around a vehicle. I slowly walked home across town. People were still wandering north. Nobody was going south. I did not carry my cellphone around then. It was more a piece of specialty gear that I brought on trips. I called my dad from the landline and told him I was okay. I got on the internet, which still worked, but I had to access it through a dialup number belonging to NYU. An email from a friend in California awaited me with the subject header "What the fuck?" It seemed as though people everywhere else knew more about what was going on than we did in New York.

I fed my cat, grabbed a camera and went out the door. I spent the day wandering around the neighborhood, watching events unfold and taking photos. In retrospect, I should have shot more than the two rolls that I took, but I was in shock. They cordoned off lower Manhattan as a gigantic crime scene and everybody seemed to congregate on West Street. We watched the emergency vehicles go south and the crushed cars being towed north. Trucks pulled crushed cars on the pavement, leaving a trail of sparks. My technologies on that day were a landline and a film camera, as well as cable television. My main connection with the world was the internet, via the dialup service that I had. Even that was dependent on having a reliable dialup number to call over the copper wire network.

I gathered with some friends that evening across the street from Saint Vincent's Hospital, where dozens of nurses and doctors stood out front in a curbside triage, waiting for the casualties that never arrived. One crazed or drunken guy charged a bunch of cops, screaming "let them die." They tackled him and led him off to a van. I went to my regular dinner place, Kate's Joint, on the Lower East Side, where I was the only customer. My friend, Ben, the only waiter, and I had a conversation that went like this:

"I don't know why I came here."

"I don't know why we're open."

I finished my food and rode back across town on my bicycle. I did not have to stop for traffic lights. There was no traffic. In the absence of information, there were wild rumors. The truth was awful enough, and many of my friends had seen people jump from the high floors of the Trade Center. There was no school for the next few days, which I spent with friends, as well as on my own, riding my bike through the deserted streets of Lower Manhattan. After a week, I went back to class and resumed my graduate education in the field of interactive telecommunications.

We had quasi-therapeutic discussions in those first few classes about what had happened. It seemed as though NYU had sent the same letter to all the professors, telling them to discuss the attack. This was probably a wise idea. You could not ignore an event like 9/11. Some students confessed to wishing that they could get a degree in a helping field, such as nursing, rather than the one we were enrolled in. Many people felt helpless.

The professor of the introductory class on coding and robotics began with "How could computational media have changed things last Tuesday?" We all looked at him blankly. It was the type of question that shows the world view of some Silicon Valley professionals. What problem exists that cannot be solved by the power of computing? Maybe it was not a stupid question, but it felt like a bad time to ask it.

52

GRADUATE SCHOOL

I BEGAN TO CARRY MY CELLPHONE MORE OFTEN. THAT was one result of 9/11. I also began to stay in touch more with my friends and family. My mom had died in 2000. The reggae band had broken up not long after that and it felt like the city was different. I was shooting with a Nikon Coolpix 995 and using my film camera at least as much as the digital. I cut back on teaching and freelance writing to focus on school.

School was exciting and predictable at the same time. People became immersed in their projects. These usually involved some combination of electronic technology and coding. Some students were trying to create the next big thing, whatever that might be. Many projects revolved around collecting the experiences of other people and then monetizing them. Some students built interesting light displays. I liked to create things with my photographs and was thrilled to learn about digital video editing.

It was at this graduate program that I met the woman I would make movies with for the next 8 years, Meredith. We made a short movie for a class project, along with a third person, some guy who had been a film student in a previous life. We shot it on MiniDV, a digital tape format, and edited the whole thing in Final Cut Pro. Even though Meredith and I came from different backgrounds – she from coding and I from writing – we were both smitten with the ease and power of non-linear editing, which is how you edit video on a computer. It allows you to manipulate your video one frame at a time with no more trouble than copying and pasting words in a word processing program.

My fellow students and I hung out and talked about technology. We started a low-brow e-zine. I seemed to create some kind of magazine any time I was in school, from the age of 10 until I was 40. My format had gone from mimeograph to Xerox machine to HTML. Everybody was hooking up with each other, going out drinking and, eventually, getting married and having kids. My only interest was the next road trip, the next camera, and doing something with my work. Nikon had a DSLR for pros, the D1, but it was too expensive for me. When they announced the D100, in 2002, I started to save my money.

By the time I entered NYU in 2001, I had been working with computers for a little over six years and made a living with digital tools. I had taught college students how to use software and how to create online content. I had built websites and staked out my tiny corner of the internet. I was taking digital photos and storing them on hard drives. I carried a simple cellphone and watched cable television. I had the same email address from six years earlier. I had picked up other email addresses, including ones from NYU and Harvard. I kept multiple hard drives with backups of all my data, as well as a schedule for creating and maintaining these backups. Modern technology was completely enmeshed with my life.

53

POWER STRIPS

I WOULD WALK HOME FROM CLASS AND THINK ABOUT modern college dorms and how different they were from what I had known. My fellow students in graduate school had ethernet and Wi-Fi connectivity. Landlines were disappearing fast. The death of the copper wire telephone network came quickly. Cellphones, initially a luxury, became indispensable and ubiquitous within a few years.

At first, people carried a cellphone in addition to their "home phone." As cell phone plans became cheaper and the network grew bigger and more reliable, households began to drop their landlines. I was at a motel in Austin, Nevada in the early 2000s and the owner, Dean, complained to me how much it cost to maintain the phone setup in his 12-room motel. For years, it had paid for itself. Now, the system was expensive overhead. People were using cellphones instead. College dorms were no different.

We all began to need more electrical outlets than ever before. My childhood room had two outlets. That house was built in the late 1940s. The techno cabin I set up in my closet at the age of eight, with its multiple adapters, three-way plugs and the toys I plugged in, turned out to be a glimpse into my future.

In college, I lived in Victorian-era buildings that were wired for electricity in the 1920s and had remained the same ever since then, with the addition of a telephone jack. Each room had two outlets. Plug in a lamp, an alarm clock, the stereo and an electric typewriter and you were maxed out. You did not need more at the time. I would unplug the small refrigerator briefly so I could use the hot pot. The outlet-to-room ratio remained constant in most of my New York apartments, all of which were early twentieth century housing. If you had two outlets in a room, you were lucky. One was usually switched, for a lamp, and the other was often inaccessible, behind a bed or the couch.

This was fine until the age of the internet, which coincided with the rise of home video games and "smart" devices. By the late 1990s, there were not enough outlets. Large power strips became the norm. My desk on Barrow Street had a double socket outlet under it and a double one by the bed. I needed power strips in both. The bed outlet fed the TV, the cable box and the VCR, along with a lamp and a humidifier.

The dual outlet by the desk saw hard use. Its interior wiring was vintage 1920s, wrapped in paper, rather than plastic, and it fed two power strips. Off these strips ran a CD player, a modem, a computer, a monitor, an answering machine, a cordless telephone, a fax machine and my first cell phone charger. I had a cassette-based transcription machine and had to unplug something else when I used it. When I began to use two computers simultaneously, the situation got worse and I had to plug one power strip into another power strip. It was dangerous, ugly and took up a lot of floor space.

In 2002, I moved ten blocks north in Greenwich Village, back to the block where I had lived when I was married. It felt as though I were moving across the country. I took advantage of the move to get broadband internet. I got a DSL line and then upgraded to faster access with my cable TV company, Time Warner. I briefly carried both the DSL and the cable service because the Time Warner connection was so unreliable. I had moved into a six-story building that was built in the post-war boom, in 1956. It had a small room in the basement, labelled on floorplans as the "telephone room." We called it the "package room," but that was where the phone company had their large access panel for the whole building.

Whenever a technician came to hook up a new line, he had to find a spot on this panel and make sure it connected to an existing line in the apartment. 2002 seemed to be the peak of usage for telephone lines. Most people still had landlines, many had fax lines, and even more had modem lines. The box was maxed out and I was worried when the Verizon technician made his visit. According to his old schematic, there was another old panel, located in the garage. He said he could connect me if he could find this panel. He found it and I got my DSL, which was slow, but reliable. I installed an Apple Airport Wi-Fi router, a white dome that looked like a flying saucer. By this time, I had bought my first Mac Pro, a gray plastic tower of a computer, and a 17" external monitor.

I set up this new apartment in direct opposition to my previous one. On Barrow Street, I had the small New York one-bedroom set-up, with a tiny kitchen in one corner of the living room. The living room was where I had my television, a couch, my books and a table. This was also the room with the air conditioner, as the bedroom had a fire escape and I could not legally install one there.

On hot summer nights, I would sleep in the living room for the air conditioning. The bedroom was where I set up my desk. As the computers and peripherals began to take over my life, the desk grew correspondingly. I added another wing to the desk after I bought my first computer. I spent most of my time in the tiny bedroom. When I wasn't working there, I was sleeping there. The living room went unused most of the time, except by the cat.

When I moved north ten blocks, I kept the bedroom for sleeping and, recognizing that I never entertained and had few visitors, I turned the living room into my work studio. I had an electrician install conduit and quad outlets along the main wall. He wired them into a grounded computer circuit. I knew that I needed many electrical outlets and did not want to fall into the trap of using endless power strips. I replaced the apartment's fuse box from 1956 with a breaker box. No more would I need to stock up on the old, screw-in glass fuses.

I got an uninterruptible power supply, essentially a car battery in a case that everything plugged into. This would give me time, in case of a blackout, to shut down my machines and save my work. It also guaranteed a steady power supply, with no spikes or dips. I built a long workstation along the main wall and was ready to go. My only regret was that I didn't have enough room to build the whole setup two feet from the wall, so that I could get behind it and run cables with dignity. Instead, I needed to get on my back and contort myself, often with a flashlight in my mouth, every time I worked on a machine or re-cabled a mixing board.

From 1960 to 2002, my living quarters had evolved from my childhood room, which contained just a few electric devices, to an apartment that was in complete service to an electronic, interconnected lifestyle. Next to my first Mac Pro tower, I installed an eMac computer for doing audio work while the big machine was busy. I had my Airport wireless network, cable television and landline telephone, along with my cellphone.

When I moved in, I dutifully set up my stereo amp, speakers and turntable, even though I rarely played vinyl anymore. I had sold thousands of LP's for drugs in the 1980s, but I continued to maintain a turntable. My cassette deck was part of this system. By now, I owned a dubbing cassette deck that had two decks in one. This made it handy to dub band rehearsals and other tapes.

I took my landline number with me, but I could only do it because I was moving within the same exchange. Now, with cellphones, people routinely take their numbers with them. They are attached to their numbers. The numbers are a connection to their original home or other location, rather than a current place. A phone number might hold memories. Friends and family have that number. There is no technological impediment now to taking your number with you.

High-cachet numbers, such as area codes in New York City, Santa Monica, or Hawaii, can now be claimed and retained. They are bought and sold through brokers. As with all fashion, though, cachet can change over time. 917 was once thought of as a lesser area code to have. It was for people who could not get a 212 number. It came to be associated with Brooklyn, which has enjoyed its own renaissance and become, in that awful term, a "brand" of its own. People can change their area codes as much as any other part of their identities. For a small group of people, they have become fashion accessories, personal statements, and trophies, rather than simply a tie to a geographic location.

My apartment was basically a large equipment closet. I had my film and digital cameras, along with countless widgets, ranging from USB hubs and Zip drives to hard drives, iPods, microcassette recorders and video cameras. I had battery-powered analog clocks and Casio watches, as well air conditioners, a plant light on pulleys, and power tools. My television was still the 19" NEC monitor that I bought in 1983. I had kept it all those years, nursing it along, needing only to replace the power supply at some point in the 1990s, back when people still repaired broken electronics.

By now, I was using hard drive enclosures that held four or five separate drives. They connected to my computer via eSATA, which was not the most common protocol, but it was fast. I could have opened an electronics store with all of the gear under, on top of, and above my workspace. I had escaped from the corporate world and created my own private cubicle.

In the kitchen were a microwave and a drip coffee maker. Both of them, along with the stovetop and oven, had digital clocks that needed to be changed twice a year for Daylight Savings Time. I also had to change the time on my answering machine, which was digital by now. I retired my last cassette-based answering machine before the year 2000, but I held on to a cassette of messages from the dark days of the 1980s.

54

MODERN LIFE

My mother had bought my father an iMac two years earlier and I enjoyed showing my him how to navigate a computer desktop and the virtual world of the internet. He found a program that taught touch-typing and he used it to teach himself how to type. He was in his late 70s by now, but he saw the value of the internet for researching things and staying in touch with people.

As a child of the Great Depression, he still loved listening to the radio. Television was his main entertainment and his method of film consumption was still VHS tapes, which he recorded with the built-in timer on his VCR. That's how most people were still watching movies in the early years of the twenty-first century. Bandwidth was slow and services like Netflix did not begin to stream video until 2007.

Netflix had been popular since 1997 for their subscription service. You signed up for a monthly subscription, got DVDs in the mail, watched them, then returned them in the US mail with the red envelopes they had arrived in. DVDs and DVD players had appeared in 1995 and were becoming popular and affordable. As the music business made their money twice over with music, first with vinyl LPs and then with CDs, Hollywood now replaced VHS tapes with DVDs.

Some people bought into the LaserDisc system. It was not easy to keep up with the pace of invention. Audio went from tapes to CDs, but there were odd formats along the way. MiniDisc and DAT machines, with their own proprietary media, could also record audio. I learned my lesson with Betamax and stuck with the popular formats after that. It was the same with digital cameras. They improved so much every year that it became expensive to keep up to date with the latest technology.

Roland and I had reunited with our original drummer and were playing the punk, rock and dub that we had played thirty years before. Little had changed, although our equipment was newer and we were older. We still loved to make music and we were better players than we had been as teenagers. We decided to record an album. In the early 1980s, we would buy studio time at dingy recording studios in downtown New York City. These places had 4-track machines and we had little money, so all we could do was throw down a few tracks, thank the engineer, then leave. We never released any of that music because we could not mix it and there was no internet to self-publish or promote our work.

Now, we had experience and some means. Roland had spent years working in audio and music production and knew what he was doing. We laid down basic tracks at a studio and he took the tracks – now digital, on a hard drive, as opposed to tape - and spent hundreds of hours mixing them. We lay down overdubs, and then he mastered the final versions. I came up with the album art, with the help of Photoshop, and we had

1000 CDs printed. We did little with CDs, but we had fun making them. We played some gigs, as well.

I still had hundreds of cassette tapes. In the late 1970s, I built a wooden box to hold my cassettes. It was about two feet wide, a foot and a half deep, and the height of a cassette case. I filled it immediately and built a crate to hold four of these boxes on rails, forming a wooden cabinet with doors. I ended up with three of these crates. Inside were all of my cassettes, from compilation tapes to albums. The three cabinets, stacked, were taller and heavier than I was. I threw out many cassettes and Betamax tapes when I moved, but I kept masters of shows and rehearsals.

55

SOUND PREFERENCES

THE INTERNET BEGAN TO MAKE MORE NOISE BY 2000. You could listen to things online before then, but it was not ideal. From the mid-1990s, RealAudio Player was an application, buggy and slow, that let you stream audio. There were two big problems. First, there was not a lot to listen to on the web. You could find some music, along with news clips and short videos, offered at tiny resolutions. Second, the connection speeds for most people were slow. Both of these issues eventually got worked out. Computer speakers became a popular accessory, especially as people began to use the optical drives in their computers to play CDs.

In early 2001, Apple released iTunes. That fall, Steve Jobs announced the iPod. I watched him introduce the product with his usual laser focus. His devotion to the product was infectious. "Your whole music library in your pocket" was the pitch. I had to have one and I bought one when it was released. The iPod felt sleek and futuristic. It would soon feel clunky and dated. Consumer electronics technology and design were about to go into hyperdrive.

The first iPod was white with a stainless-steel rear housing. It had a backlit LCD screen on the front, monochromatic black on gray. The controls were elegant, but they would soon feel old-fashioned. The main control was a scroll wheel whose outer ring was made of four clickable buttons. The inside of the wheel was a plastic circle that moved as you turned it. This would become a touch control in the next version but, on this first one, you could rotate the plastic wheel and physically click the outer buttons.

It had a "Hold" switch that locked all the controls if you wanted to stick it in your pocket and not worry about pressing any buttons. This scroll wheel iPod used Firewire to charge and connect with a computer. It had a capacity of 5 gigabytes, which was huge at the time. It came with a set of ear buds, which never stayed in my ears, leading me to conclude that my ears are shaped differently from those of other people.

I began to transfer my CDs into iTunes on my computer. This process, grabbing analog audio or video and converting it to digital, is called "ripping." Of course, the more I ripped, the more storage space I used. Hard drives were still expensive. They were a hassle to install inside the Apple computers of that generation, so I was frugal with the bit rate, or the quality level at which I ripped this music. I was choosing quantity over quality. My descent into Lo-fi sensibility had begun. The difference was barely noticeable on those shitty computer speakers.

Much more compelling than the quality was the ability to organize and play music in a completely new way on the iPod. Digital music players became smaller and cheaper every year. You could play an endless stream of music. No longer were you restricted to an album side or a cassette's worth of music.

You were not limited to a stack of 7 records on the changer. You could play songs for as long as you could listen.

This was good and bad. There is something concrete about listening to an album side. It goes in a linear way from start to finish, and artists had been designing them this way since the Beatles. Before then, albums had been excuses for showcasing one or two hit singles, with the rest of the material being fluff and lesser work. By the time I began to buy records, in the early 1970s, albums had become unified concepts. You would bring one home and rip off the shrink wrap, hold it in your hands and then put it on the turntable. You would usually listen to it from start to finish while examining the album cover and reading the liner notes.

It was a holistic experience. People discussed the merits of "Side A" versus "Side B" or, in the case of a double album the "third" and "fourth" sides. Roland and I, through some accident involving skill and obsession, had memorized the guitar parts to the third side of Jimi Hendrix's "Electric Ladyland" and played along with that whole side over and over in his basement when we were teenagers. Certain albums were known for their psychoactive arc. They would build and peak and then let you down slowly.

Critical discussions focused on whether or not a group should have released a double album, whether or not they had enough material, and whether the consumer should be forced to pay for a lot of crap in order to buy the good stuff. The Clash, famously, made "Sandinista" a three-album set and took a financial hit, against the wishes of their label, because they wanted to give something back to their fans. The long-playing album become the unit by which music was measured. People bought singles, but most people bought albums. Radio stations dealt in singles, but singles pushed album sales. The internet and the iPod would change all of this.

I ripped the CDs into my computer and then synced my iPod, filling it with music. I began to go out wearing headphones. I would walk around and let the songs shuffle - play in random order - and enjoy the segues that the software generated. My DJ partner from college radio and I had sat around, smoking many joints and discussing the art of the segue, so I had to accept the idea of segues being generated by machine. As with digital cameras, the original iPod became obsolete before it got old. The next model was bigger and better. It was smaller, but it had more capacity. The interface was still clumsy, but they would fix that on future versions. I bought a tiny iPod Shuffle, as well.

When I played music live, I still taped the shows and rehearsals with a small cassette deck, but now I began to digitize the tapes. I wanted to digitize a lot of other cassettes that I owned, as well. Audio cassettes had been a reliable way for me to store music for years, but I was starting to worry about them. Recording tape is made by coating plastic with an oxide that can be magnetized. This oxide eventually starts to flake off. Converting tapes to digital bits and bytes seemed like a good way to arrest the decay. Of course, there are problems with digital media, but I wasn't worried about that yet.

I began to conduct experiments and do research on the best ways to digitize cassettes. By now, "doing research" on almost anything meant searching the web, rather than consulting books or talking to other people. If you could not find a web page with specific instructions for doing something, you could always find a discussion board somewhere with the information. Some people were getting a lot of traffic on their blogs and starting to monetize those clicks and discussions. Many useful technical websites and online communities were born at this time. I wasted many happy hours on them.

In 2003, Digidesign released the MBox and the Digi 002. These two hardware interfaces, combined with the ProTools software, looked like the best way to convert my analog collection to digital. It was not what ProTools was designed to do, but it seemed solid and future-proof. I began to sit around in my spare time and feed cassettes into the system that I created. I wound up with digital audio files on my computer, but I kept the cassettes, just in case of a catastrophic failure.

56

13BIT PRODUCTIONS

2003 WAS WHEN I GOT MY MASTER'S DEGREE FROM THE Interactive Telecommunications Program. My thesis consisted of a paper and an interactive DVD that worked with a live website. The content was my writing and photography. After graduation, I began a creative partnership with my classmate, Meredith. We were curious about the software that the new generation of VJs was using. These VJs - video jockeys and video artists – were doing fascinating work. They used software and old, re-purposed analog equipment, to make visual art, usually projected, for clubs and galleries.

Meredith and I had worked well together on our short film in school and thought about documenting what these video artists were doing. After graduation, we began to shoot interviews with them. Documentaries were becoming cheaper to make than ever before. We shot no film. We used a Sony PD-150 video camera and edited in Final Cut Pro. We bought lights and microphones, along with a tripod, and we were ready to go. We called ourselves "13BIT Productions," which was our little joke about how things work. In the computer world, you always have an even number of bits – 8-bit music, 16-bit applications or 32-bit integers. We liked the idea of 13 bits. Also, the domain "13bit.com" was available. We started our empire.

We taught ourselves lighting and sound, as well as color correction. We enjoyed the process. The documentary became as much about the VJs as it was about the technology and the art. It was liberating to make the whole movie on a desktop computer. We named it "Video Out," a title based on the output jack for a video signal. We entered it in film festivals and won a few awards. This film led us to make another documentary, the precursor to "Video Out."

This second movie was about Thomas Wilfred, an inventor and artist who created projected light art with electro-mechanical and optical machines in the early part of the twentieth century. The documentary was about art and technology, as well as Thomas Wilfred. We called it "Lumia" and it did even better critically and on the festival circuit. Wilfred was ahead of his time with technology and aesthetic. The older light artists had mentioned him to us as an inspiration. If he had lived ten years longer, he might have gotten more acclaim in his lifetime. As we cobbled the film together in Final Cut, we realized that we had a movie about artistic and technological tragedy on our hands.

Meredith and I spent whole days happily editing video. We became masters of video compression, color correction, and audio design. The technology was a challenge, not a chore. We entered film festivals and were featured in many of them. At the time, you needed to send a film festival entry as a DVD or a VHS tape, along with the paper entry package. This meant going to the post office. I loved taking the subway up to the main branch of the US Post Office, on 33rd Street, late at night. Soon, festivals would only allow you to upload your entries. Physical media was dying.

As side projects, we made podcasts. We created blogs. We made a website about low-budget filmmakers. We hosted a robot parade for a few years in Tompkins Square Park. That parade was more about the costumes than actual robotics, but our heart was with robots everywhere. We may not have had a business plan, but we did have fun.

57

WAITING FOR THE MESSIAH

In 2004, friends of mine began to get invites to Gmail. Whether Google was limiting the volume of requests or it was simply a clever marketing device, it was effective. Everybody wanted an invite. I wanted an invite. Who wants to be left out? At the same time, I was happy with my local mail. I did not want my emails sitting on some distant server. I preferred to store it on my computer, rather than in the cloud.

I had begun to question what Google was doing with my search data. I did not want them to read or search my email. When I finally did get an invite, I started an account, but used it for very little. I set up an auto-forward, so that any mail I got on Gmail would end up in my email program. Most other people I knew dove in with both feet. Gmail was fast, relatively reliable, accessible from anywhere, easily searchable, and free. Very quickly, Gmail came to be synonymous with "email."

Meredith and I stayed busy with filmmaking. I was buying more equipment and spending more time learning how to use it. The technology was moving so quickly that gear would become obsolete before it was a year old. I spent a lot of money on an adaptor made by the Letus Company. It let you use SLR lenses on the Sony EX-1 video camera. It was expensive, imperfect and heavy, making it impossible to do any handheld shooting. It gave our video footage a more filmlike look. Within two years, DSLR cameras could shoot video and the Letus was obsolete. I sold it for a loss. We hired a sound person to capture audio on our movies and he arrived with a trunk of gear. It felt like my life was made of electronic devices and cables.

In 2005, I traded my Jeep Wagoneer for a Toyota Tacoma pickup truck. I had nursed that Jeep along for years, but it had turned into a money pit. More ominously, the engine was starting make odd noises. I was tired of breaking down in strange places. A Toyota dealer near me offered a trade-in on any vehicle you could roll into the lot, working or not, so I took advantage of it and bought a truck with four-wheel drive and a locking differential.

I bought a fiberglass cap for the bed and installed a heavy steel bumper. This offset the balance, so I had to add two heavy-duty coil springs to the front and two leaves to the springs in the rear. I bought a second spare tire, just in case the first one blew out and I was far from civilization. When I was done, I had a vehicle that could go anywhere I wanted it to go. I felt sad that the Wagoneer probably ended up as scrap.

The Tacoma was a joy in terms of technology. Nothing in it was cutting edge, but it was the first modern vehicle I ever owned. I finally had air bags, a good sound system, oil that did not need to be changed every 3000 miles, and fuel injection instead of a carburetor. The engine block was not coated with grime and carbon. There was no vacuum system to spring leaks. The little electrical switches all worked and the engine, with two fewer cylinders than the AMC 360 V8 in the Jeep, felt strong. Best of all, it had a new car warranty.

As I got older, I began to appreciate warranties more and more. With technology and computers, I usually did not buy extended warranties, but with high value stuff, I did. Of course, I got ripped off by the car dealer, but I was naïve. The Tacoma had a couple of power ports – they were already being used for electronic devices more than for cigarette lighters, but they kept the same form factor. The Tacoma also had more wiring and microchips in it than the Wagoneer.

When I was younger and we talked about cars, the common wisdom went that all cars after 1974 sucked. This was because of the "pollution controls" they added to comply with emissions standards after the gasoline crisis that year. That may or may not have been accurate, but it sounded good and made whoever said it seem as though they knew what they were talking about. I honestly didn't know what was going on with the electronics in that Tacoma, but I do know that it felt like the Lunar Landing Module to the Wagoneer's Graf Zeppelin.

I was still teaching in 2005, mainly Photoshop classes, but also the "Politics and Mass Media" class at the New School. I took road trips as often as possible. There were some incremental changes to the technology in my life, aside from what I used in video production. Computers were getting more powerful, as were digital cameras. I upgraded as often as I could afford. Software was becoming more stable, but more bloated. Software took up more room, as did the files that I created.

My first computer's hard drive was 345 megabytes. That was generous in 1995. Ten years later, I was buying hard drives that had capacities in gigabytes. The music, photo and video files were getting bigger. Soon, I would be using hard drives that measured in terabytes. Digital camera sensors were getting bigger, as well. It seemed as though Moore's Law remained in effect. I was using a clamshell phone with a tiny mechanical keyboard.

Text messages had been around since the early 1990s, but nobody I knew was texting until 2000, and not in a serious way for a few years after that. I was not keen on using my thumbs to type messages, so I rarely texted. When texting became pervasive, my relationships with people began to change. Texting caused communication problems. With voice calls, you lose the nuance in face and body language. With texting, you lose the voice, as well. It's easy to misunderstand a person when they speak in short messages. People began to use simple emoticons. :-) for a smile, ;-) for a wink, and others. These helped to convey intent.

With the first generation of cellphones, text messages were not grouped as conversations. You got a text, then you might get one from someone else below it. It was easy to lose track of what was being said. It was not intuitive. If somebody sent a photo from the low-resolution camera on those early phones, you would get a text saying something like "I sent you a multimedia message. You can view my message w/in the next 7 days via the web at www.viewmymessage.com using MSG ID b0dsanml Password wear4ddat." This was a hassle. The companies were developing technical protocols, but the user interface was terrible. I had a series of Nokia and LG phones. They steadily improved, but nothing felt earth-shattering. I used my cellphones mainly for telephone calls. Texting felt like a cheap trick.

In 2007, Apple announced the iPhone. It would go on sale June 29th and - among Apple true believers, at least - the excitement began to build. Many media pundits and well-known technology writers predicted it would be a failure. Prognosticators said that Blackberry, Motorola and Nokia had nothing to worry about and that the iPhone would appeal to only a small number of gearheads. Microsoft's Steve Balmer laughed at it on record. I knew I would buy one. I placed a pre-order as soon as it was possible and then I waited.

Of course, I didn't just wait. I read everything I could on every technology blog and news site. By this time, I began my day with technology blogs, along with my morning coffee. These sites were the first places to post troubleshooting articles, to identify problems and to share solutions. They even provided low-level social interaction among their users. Information would dribble out from anonymous people in the supply chain, from "well-connected insiders," stock analysts and other sources. We heard about bugs being worked out. After the presentation by Steve Jobs, there was a lot to speculate about.

58

IPHONE

I GOT MY IPHONE VIA FEDEX ON THE DAY IT WAS released. I lifted the cover off the box and got to the phone inside. I marveled at how solid it felt in my hand, how sharp the screen looked, and how intuitive the user experience felt. I was holding a tiny computer. I experienced another one of those moments, an instant recognition of technological progress. It took a while to activate the iPhone over the network. The servers were overwhelmed.

When it was finally up and running, I made a few calls. I started the web browser and began to hit some of my regular web sites. When I saw them displayed on that handheld screen, it was a moment of pure possibility. It was wondrous to hold a tiny web browser in the palm of my hand. In my emails that day, the only one that mentions this device was to my business partner, Meredith. The subject line was "iPhone" and the complete email was "It has a removeable sim card," so we must have been talking about that. I found Apple's marketing slogan, "The internet In Your Pocket," to be accurate. This feature was a revelation at the time.

Six days later, other emails appear in my outbox, from me to a bunch of Craigslist ads selling iPhones, with me saying "Do you still have it?" I had gone to the movie theater on Sixth Avenue and Waverly Place to see "The Future is Unwritten," about The Clash's Joe Strummer. I settled into my seat, but the seat bottom would not tilt all the way down. I used my butt and my full weight, caveman style, to pound the seat down. It would not completely settle in, so I got up and looked down through the crack at the back of the seat.

I could see a blue glow on the floor and had a sickening moment of awareness. My new iPhone had fallen from my pocket and I had smashed it to pieces. It was completely bent in half, still glowing, but far beyond any hope of repair. It was just a twisted hunk of metal and glass. I got on Craigslist and found someone who was not a speculator, but who had simply bought two phones from different sources in the hope of getting one. I paid him cash and was in business again. After six days, I could not live without the iPhone.

Some people were calling it "The Jesus Phone" or "the God Phone." Apple made an ad playing on that concept, with a finger reaching out toward its lit face. I used the email, web browser, calculator and other functions immediately. I lay in bed one night that first week and watched a Jimi Hendrix clip on YouTube in the dark. It was miraculous – handheld, wireless internet. I gave up my plan with Verizon so I could buy this phone. You could only get it with AT&T at first. It soon replaced my iPod. The camera was truly useful, as opposed to the cameras in all the other cellphones I had ever owned.

The way it handled text messages was also a breakthrough, and text conversations now made sense, as they were grouped individually. The onscreen keyboard worked well. Any deficiencies the phone had were outweighed by the brilliance of its design and what it delivered. At first, there was no App Store. Seeing web pages rendered on the iPhone's browser gave me a thrill. Initially, you saw the same web site that you would see on your computer, rather than "Mobile Sites," or versions that were streamlined for phones. These would appear later.

I began to notice other people on the street with iPhones. You could identify them by their lowered heads and intense concentration on the screen. The exterior world had melted away for them. They were focused on whatever was on the screen, whether it was the web, a text conversation or email. What previously had taken place only in front of a computer was now happening everywhere.

It was a deep connection, this bond between user and phone, and it began to organically supersede the human-to-human connection. Meredith and I were shooting a movie at the time and she was re-negotiating her mortgage. There were two days when she was absorbed in her iPhone on set and nothing was going to get between her and that device. In the past, you might have excused yourself for half a day, taken care of business, and then returned to whatever you were doing. Now, your business went with you, and it was often shared with everyone around you, whether they liked it or not.

True mobile computing was exciting at first. Laptops had once been revolutionary, especially when Wi-Fi became common. The iPhone was a computer in your pocket. It was networked, as well. I was with my friend Paul that fall, driving east toward the Donner Pass on our way to soak in some remote Nevada hot springs. I had spilled coffee on a shirt at breakfast. I did not need to replace it immediately. In the past, it would have waited. I pulled out my new iPhone, found the shirt on a website, and placed the order, pinching and zooming the screen to press the tiny buttons, before we left the cellular service zone. I didn't have to do it, but it was a fun and radical at the time, so I did.

Listening to random, anonymous banal conversations from other people on the street did not start with the iPhone, but it certainly accelerated. More people were buying smartphones and becoming louder in public. Many wore headphones or Bluetooth headsets and walked down the streets, screaming into their phones to overcome the ambient noise. A friend's mom, a therapist since the 1970s, said to me "It used to be that you knew who was talking to themselves. Now, you can't tell." The tipping point had arrived. Within a year, I got onto a subway car and the majority of people – instead of reading papers or books, sleeping, or staring – were all stuck, face-first, in their smart phones.

My first iPhone had 16 gigabytes of memory, almost 50 times as much storage as the hard drive on my first computer. It did everything I could have asked for and more. It was a revelation, then it became a necessity. My geeky friends and I were focused on what would come next and how Apple would improve it. The other phone makers were scrambling. Another high-tech race had begun.

Within a year, Apple announced the iPhone 3G, named for the 3G, or "third generation" network technology. It had a bunch of enhancements and, as Apple said, it was "Twice as fast for half the price." Whether that was true or not was immaterial to me. The big news was the new App Store. It did not seem like it would change my life, but I was open to the possibility.

The first app I saw anybody buy was called "iFart," which an obnoxious acquaintance installed and gleefully used everywhere. Predictably, it made farting noises and was on par with the whoopee cushions that were advertised in comic books in the 1960s. It was a party novelty. In July of 2008, I installed the app for eBay. A few days later, I bought an app for the New York City transit system, along with phrase books for Chinese, Japanese and Vietnamese. I must have wanted to travel to Asia, but I didn't go any further than buying the apps.

Apps were cheap, so it was like buying a lottery ticket – "it's only two bucks." I bought apps for the weather, navigation, photo manipulation, meditation, barcode scanning, audio creation and others. Nobody realized, at first, how profoundly the App Store and the app economy would change our lives. They were free or cheap. Buying and using them was intuitive. My phone was now in my hand and my face all the time. The cell networks were getting more reliable and data plans more generous. I had an unlimited data plan with AT&T. They tried to eliminate those plans, but some people held onto them for years.

Between the better networks and the cheaper data, the smartphone became the ultimate tool to have on a road trip. You could use it to navigate, to check on motel rooms and ratings, to take photos and to stay in touch with friends via email and text. You could even make phone calls with it. The large parts of the map that had no cell coverage became more frustrating to endure. "Addiction" was too strong a term, but "dependence" may not have been. I was using it all the time. Only ten years earlier, I had cobbled a wonky internet setup for the road with a laptop, digital camera, cellphone and cell modem.

I preferred a physical road atlas. The last thing I do each night in a motel is get into bed and read the road atlas. I like to scout ahead. You can get the big picture more easily on a road atlas than with a cell phone's mapping app. Still, the iPhone removed the chore of adding up mileage segments in my head and calculating how long it would take to arrive somewhere. It would eventually bring live traffic data into the equation.

I followed the online war between Apple Maps and Google Maps. Most people seemed to favor Google Maps. Whether this was because they were familiar with it or, as I darkly suspected, there was an active, covert smear campaign against Apple Maps, we will never know. I thought Apple Maps had been unfairly maligned. Google may have had more data to back it up, but I preferred Apple's interface. It was like the VHS/Betamax war all over and I once again chose what seemed to be the losing side.

By this time, a common epithet you could hurl at a fellow technology lover on an internet forum would be "fanboy," whose real meaning was to have a blind allegiance to a company or product, no matter what faults it had. I was clearly an Apple fanboy, but I was not opposed to other mapping applications. When the iPhone came out, I paid more for the TomTom and Navigon apps – about thirty bucks each – than I would ever pay for any app ever again. I did not like either of them, but I was hoping one would work out. As Apple's iOS mobile operating system kept improving, I stuck with Apple Maps and never looked back.

59

FACEBOOK

I JOINED FACEBOOK ON AUGUST 6, 2007. MY "Registration Confirmation" is the first email in my "Facebook" mailbox. I created this mailbox because Facebook sent an email whenever they announced anything. I could have just deleted those emails, but I saved them. I first heard about Facebook a year or two earlier, while teaching my students at the Brooklyn Campus of Long Island University. I overheard a few of them saying the name "Facebook" and making it sound like a physical place, some club they went to. I made a mental note to check it out.

I don't remember why I joined, aside from that primal desire for community. I had belonged to AOL and even MySpace, briefly, so maybe I was just curious. The next day, I joined the "Harvard" network, as they called it on Facebook. I have an email from an actual human, confirming that he made the changes I requested to my Facebook profile. Following that in my mailbox are hundreds of emails, informing me that I had added people as friends or made comments about things.

Some emails suggested that I become a fan of something or that somebody had sent me a message.

At first, I felt that I had found a new community similar to the one I had lost on AOL. Friends were here and I was again in contact with people who had dropped out of my life for decades. In the beginning, my posts were more personal and confessional. Over time, I revealed less and less of myself. I found Facebook a useful place to post photos and get feedback, though not necessarily overt feedback. I would see what photos people liked. Some of them surprised me. That was useful.

I decided early on that I did not want to be "friends" on Facebook with anybody I didn't know personally or had not met at least once in real life. I am not a people collector. Over the years, the list of people I have blocked on Facebook has gotten longer than my list of friends. As the country became more polarized politically, I began to block anyone who expressed extreme positions, even if they were a complete stranger. I did not want to associate with people who didn't agree with me. That's a terrible attitude for the greater good and society, but that's where I ended up. I would not want to talk with a boor spouting lies and vile opinions at a party. Why should I on Facebook?

A friend of mine, on first stepping into an AOL chatroom in 1995, said "It's like the great American cocktail party." A cocktail party is physical, though. There is no established set of norms or etiquette to deal with some of the situations that technology has foisted upon us. Facebook began to make their privacy settings more obtuse and difficult to access. People began to learn how to navigate online social encounters, but still said things that they would never say in real life, face to face. Still, the network was handy to have around. I reacquainted myself with my Latin teacher from high school, a friend from my elementary school back in Yonkers, and people I had worked with over the years at various jobs.

I found myself venting about political, environmental and quality-of-life issues. I found myself venting all the time. This could take the form of rants or simply posting memes – photos with accusatory captions. It was easy to complain about anything. New concepts arose around online human behavior. "Virtue signaling" meant that you were trying to present yourself as virtuous and mindful of the greater good, whatever that happened to be in the social group you ran with. "Image crafting" was a more general term that described how you formed a self for others to see, whether that was true or not in real life.

Propagandists and political activists recognized the power in a system where like-minded people congregated and could easily share information and create an echo chamber. "Social networks" were perfect places to create "alternative facts" and false narratives. There was no real policing or regulation and the norms that had previously governed journalism disappeared, yet the influence remained.

In the past, we had all gotten our facts from the same news sources and could argue and debate from these same facts. Now, we argued about what the facts were. People began to post headlines from websites across the spectrum and others would accept them as true. There was no vetting involved. A study at this time said that most people who re-posted articles had not even read past the headline.

I became distrustful of all big data-driven companies. In 2009, I began to look for alternatives to Google. I tried Dogpile and

Bing. I never went back to Altavista or Hotbot. Like an old love affair, I could not reignite the feelings I had for those search engines. I settled on DuckDuckGo. I was suspicious that DuckDuckGo was returning worse results than I got through Google, so I would conduct side-by-side checks.

Google almost lured me back into their fold with Google Street View, a brilliant invention, made possible by their obsession with indexing the planet and also by their deep pockets. Few companies could send out fleets of cars equipped with multiple cameras and document every road in the country. In every other respect, though, I began to stay away from Google by this time.

60

THE VAN

In 2009, I traded in my Toyota Tacoma and bought a Ford E-350, a white cargo van. I had a grand plan to get a van, convert it to four-wheel drive, then build it into a tiny camper. I bought it from a Ford dealer in Portland, Oregon, who delivered it to a conversion shop on the other side of town.

The conversion shop installed the axles and running gear from a 2009 Ford Harley Davidson pickup truck onto the Van, swapped out the gas tank for a 46-gallon tank, put in a geared limited slip differential in the back and an electric locker on the front differential. I remembered my dad explaining to me – using both hands around an imaginary ball – how a differential worked when I was a kid. He explained to me, as only an engineer father could, how it was necessary for the wheels to spin at different speeds when turning.

The upgrades I made to the van were strictly for off-road use. I added a large roof rack, along with a heavy rear bumper and a massive deer bumper on the front. Ever since my friend Paul and I had totaled his Isuzu Trooper, hitting a deer in the

Nevada outback in 2002, I was a fan of heavy-duty bumpers. I set all this in motion, then packed up the Tacoma, whose trade-in was part of the deal, and left town with my dog, Elko. He had joined me by leaping into the Tacoma three years earlier in the middle of the Great Basin Desert.

We set off in late December. The light was weak, so I did not take many photos along the way. We spent Christmas Eve at a Motel 6 in Las Cruces, New Mexico, and arrived in Portland a few days later. The van was delayed, so we wandered around Oregon every day until it was ready, then traded in the Tacoma, transferred all the gear, unsecured, into the back of the van, and set off for home.

Elko sat on top of that pile of stuff – sleeping bags, milk crates filled with food, tools, and clothing, a spare tire and an off-road jack, in that metal box of a van - and stared at me up front the whole ride home. I'm sure he missed the comforts of the Tacoma. I had gotten a heavy-duty alternator installed in the van, as well, but held onto the original one, just in case, having picked up the habit of carrying spare parts with me at all times when I owned the Jeep Wagoneer. I eventually was carrying the alternator, a fuel pump, spare headlights and a serpentine belt, along with smaller parts, at all times in the back of the van.

I had been trolling the "home-build" sections of custom van and RV sites for months, so I had a good idea of what I wanted to install when I got home. I don't do electrical work, so I found a professional in Central Jersey who did marine and RV electric installations. He had been in the Air Force and was brilliant at what he did. He installed some deep-cycle batteries, the kind that you find in golf carts. These fed a large AC inverter, which powered four quad outlets so I could plug in appliances. He put a solar panel on the roof and built a control panel to monitor everything. I now had electricity for multiple devices on the road, as well as the power tools I was using to finish the van.

I screwed half-inch plywood into the floor over insulation, stuck sound-deadening panels all over the interior, and smeared a thick coat of "sound-deadening sludge" from a five-gallon bucket everywhere. It dried to a bumpy matte black finish. I sprayed thick adhesive on the wall and lined the whole van with foil-backed jute. I covered the jute with pure aluminum tape, then tried to cover the tape with birch veneer. My wood bending skills were weak, so I stopped at the ceiling. Next came a layer of radiant silver fiberglass tape, but I became paranoid about inhaling fiberglass, so I sealed the whole thing with more aluminum tape. I built shelves to the contour of the van walls, constructed a platform with storage space underneath and put a futon on top. I was ready for my first road trip in the van.

In May of 2010, I called a stop to any more van improvements, loaded up on food, water, a duffel bag of clothes, dog food, photo gear, a laptop bag and assorted peripherals. I carried firewire cables for my spare hard drives, 30-pin cables for my iPhone and iPod, and power cable for my computer. I brought a Compact Flash memory card reader for the cameras, along with Nikon camera batteries and an extra battery charger. By this time, flying anywhere or driving on a road trip required multiple power supply units, car cigarette lighter adapters, and accessories. International travel called for even more adapters. I began to fantasize about taking a trip with just my iPhone, which would act as phone, computer, and camera. It was the most minimalist thing I could imagine.

Elko and I took off on a relatively short cross-country trip. It was more of a test run than my usual photographic odyssey. We made it to the Pacific and back, sleeping in the van half of the time. The electrical systems were a success. I found myself sitting on the edge of the sleeping platform in truck stops, on beaches and in national parks, checking my email and processing photos. Elko had a small jump seat behind the driver's seat. The Engel refrigerator, with its high-efficiency rotary compressor, made a pleasant hum at night. I had small computer fans to move the air around and homemade foil panels to cover the windows. It felt as though I were in my own lunar command module.

I used my iPhone extensively on the that trip, for both data and calling. Most cell phone carriers had dropped roaming plans by then. Expensive long-distance calls were a thing of the past. Still, we were paying a lot of money every month on cell-phones with audio quality that was far worse than landlines. Convenience and ubiquity won out over clarity and most people were happy to accept the trade-off. Older people in my building would tell me "I'm never giving up my landline," but Verizon had already vowed to dismantle the copper wire network in favor of their fiber-optic network. The future steamrolled ahead.

61

POSTMODERN LIFE

I RETURNED FROM THAT TRIP MORE FOCUSED THAN EVER on photography. I had to keep up with all the latest hardware and software. New cameras appeared every year. Old software became obsolete every year and needed to be replaced. Cameras, software, and desktop computers were expensive, but external storage was getting cheaper. More people relied on the cloud to back up their machines, send files and use as a shared space where they could collaborate with other people. The line between physical and virtual was getting blurrier by the day.

Technology crept into every corner of my life, from the digital white noise generator that I used to sleep to the automated coffee maker that I used to wake up. Cars began to feature screens on their dashboards. Transit systems released apps to track your bus and subway in real time. My digital cameras became more capable ever year, as did video cameras. YouTube began to support 4K, high resolution, videos in 2010.

The internet had grown from a strange landscape with interesting, quirky web sites into an expanding universe of information. In 1995, I had worried that people were going to figure out how make money on the internet and ruin it. By 2010, many sites were behind a paywall. Pop-up ads were pervasive. Ad blockers were a niche industry. Online scams, blackmail, and identity theft had also become common. It was more lucrative to steal someone's information and identity than it was to mug them.

There was no longer a line between my life and the internet or online services. Everything was now mediated through a digital network. Rather than watch a baseball game live on television, I would go to a website that showed a baseball diamond and simple drawings of players, along with a listing of balls and strikes. I could have my phone ping me or vibrate when important news broke.

Live sporting events and concerts featured massive screens so fans could watch enhanced versions of the real life in front of them. This was not the same as the closed-circuit theater broadcast that my dad and brother took me to, in Washington Heights, in 1974, to see the Ali vs. Frazier rematch. It was not equivalent to the crowds that used to gather around the Times Square electronic "Zipper" ticker that used to broadcast news in real time.

News articles had comment sections by now and you could read what your fellow citizens said about any given story. This was usually depressing. People were paid to "moderate" these comment sections at some of the well-heeled news outlets. Everybody had something to say and many people were saying things to no one in particular.

By my late forties, I was watching less TV than ever. I had a small pile of remote-control units. They cluttered up a tabletop and often got misplaced. A remote for the TV, for the cable box, for the Apple TV, for the old VCR, for the DVD player I no longer used, and even for the DV videotape editing deck. I had a small remote-control unit for my air conditioner and a primitive one for the ceiling fan in the bedroom. I needed all of these oblong plastic boxes, each with a bunch of tiny plastic buttons that squished when you pressed them. Some of my devices would not function without them. It was clear to me that people were less sedentary in my youth, when you had to stand up to do anything and when you had to leave your house to shop.

I was paying for a premium cable package that included a DVR, or "digital video recorder," but I watched little except for the Weather Channel. The Weather Channel had been a guilty pleasure of mine since the mid-1980s. It was not just about the weather. There was something comforting about hearing it in the background. I would gaze at the strange news that only they posted and try to cipher where the humanity was behind those tiny weathercasters with the permanent smiles.

When they changed their local forecast from fuzzy, generated characters with cheesy instrumental music to slick graphics and a robot voice reading the news aloud, I was angry. I was also powerless. I had always been powerless. If public relations had taught me nothing else, I learned that everything on TV was advertising.

Apple released the iPad in 2010. I bought one. Developers created new apps for the iPad and other tablets. I bought music apps, DJ'ing apps, weather apps, photo manipulation apps and more. I barely used some of them, but I loved the form factor of a large tablet. My ultimate use for the iPad, though, was reading in bed. I could fire up the web browser and read the depressing headlines as I fell asleep or when I woke up. I could check email and online forums. I could read books. It changed the way I consumed media and lived my life. Why sit in front of a computer when I could get into bed and read? Amazon released the Kindle app for the iPhone and iPad in 2010 and I began to buy e-books.

E-books did not have the sensory impact of a book, nor did they read as elegantly. I had a nagging resentment that they would disappear after I died, although I didn't care so much about that. I could carry a bookshelf onto an airplane, though, and I could jump from book to book without moving, from the comfort of my bed. I bought a Kindle 2 when it was released, but barely used it.

The Kindle's screen quality was terrible. The text looked like it was printed with an old Etch-A-Sketch toy, the user interface and the Wi-Fi technology were primitive. It never pulled me in. The Kindle app on the iPad, though, looked great. I still preferred the smell and feel of paper, but books on the iPad were too convenient and became the ultimate way for me to read while traveling. Being able to click and hold on a word for the definition was also a little magical.

Steve Jobs died in 2011 and the media was quick to point out that people got the news of his death on devices that he had invented. In less than four years, smartphones had become ubiquitous and essential to a majority of people in the developed world. Articles from 10 years earlier, where the pundits had proclaimed "the cellphone is the new cigarette," seemed naïve and outdated. Smartphones, soon to be called "phones," were becoming as powerful as computers. Many people discovered they could fill all their needs with a phone instead of a computer.

Operating systems were constantly being updated on computers and portable devices. This was to introduce new features and protect against the endless security threats. From my first days of computer ownership, I had felt a thrill to update my devices, install the latest software, and move to the newest version. Sometimes, I paid the price for being an early adapter. By now, the updates felt like a chore. Software engineers worked more on notifications and how to get a user's attention. Life became a series of pings, bleeps and illuminated dots. More things distracted me from physical life and invaded my consciousness.

62

THE APP ECONOMY

My work with 13BIT Productions ended in 2011. We had made two successful documentaries and two experimental narratives. We had learned a lot and developed a good working relationship, but we could not devote enough time to making a film production company work on the backs of two people. We ended things after shooting the footage for one last documentary, which remains in the can to this day. Not only is the content aging, but the video format we shot it on has become obsolete. We may edit it one day. If we do, we'll have to learn a new non-linear editor, since the original Final Cut Pro is obsolete, as well.

In mid-2011, Uber started doing business in New York City. You would download the app, enter your information, and a car picked you up. I began to notice people standing on the street, staring into their phone as they watched the tiny representation of a car approach them. Hailing a cab used to mean stepping into the street, a street you chose strategically, based on the odds of finding a cab. You would look in all directions and search for a vacant taxi. You would frantically wave and hope that the driver spotted you. Until 2007, taxis took only cash. The number of cabs on the street was strictly limited.

Now, people called up Uber, Lyft and Via rides on their phones. As the use of Uber exploded, city traffic came to a standstill. A hundred thousand new cars created the worst traffic congestion ever in New York City, but people loved the convenience of summoning a car. My quiet little block became an Uber thoroughfare. I parked outside an ice cream shop on West 10th Street one day and a woman immediately ran to the rear, opened the hatch and tossed her suitcase in. I yelled back at her that I was not an Uber. She apologized, then began to scan her phone again, zombie-like, to see where her actual car was at that moment.

Grubhub, Seamless and other delivery apps sprang up and allowed people to summon food, clean their clothes or buy anything from anywhere, courtesy of underpaid gig economy slaves on electric bicycles. Takeout food once meant pulling a stack of stained paper menus from the shelf, going through them, maybe even making a list on paper, then calling the restaurant. You placed the order and asked for a total so you could be ready when the delivery came. When the delivery person arrived, you handed them the money, plus a tip.

By 2011, I could invoke an app or hit a website, place the order and pay for it, including the tip. The delivery person would arrive and hand me a bag, often wordlessly. You could now get anything delivered quickly and easily with minimal human contact. It began to feel as though the tech companies were trying to get rid of humans as much as possible. Humans cost money. And, though it took a while to sink in, human contact was becoming a luxury commodity, from customer service to sales.

Everything became on-demand. I bought an app for faxing. If I wanted to send a fax in the 1980s, I would walk to the corner greeting card and stationary store, hand the person a dollar, and they would feed the fax into a machine and give me a paper receipt. I bought a fax machine in the 1990s. When the app economy arrived, I tossed it, bought the app and signed up for a service that gave me a virtual fax number. I rarely fax now, but it let me throw out another redundant machine.

I could now lie in bed and fly around the map with real estate apps on my iPad, dreaming of places to move. I could register with an online dating app, write a profile, feel foolish, then sit back and wait for nothing to happen. I could plan trips and buy my tickets online or on the airline's app, reserve a car and get a hotel, all without ever talking to another human. I could quickly locate the next highway rest area with an app.

The media were beginning to put out app versions of their publications. The New York Times began to charge for content in 2011. This did not strike me as unreasonable. I always paid for my newspapers until the web came along. For some people, though, who expected all news and media to be free, it was an imposition. Advertising revenue was falling fast at paper-based publications.

The Times had introduced a premium section, called "Times Select," in 2005, but they discontinued it after two years. They experimented with other payment models and settled on a subscription method that let people read a few articles for free. Heavy users had to subscribe. Advertising money was flowing to the web, along with annoying pop-up ads that took up half the screen. The news and entertainment industries had to adapt or go under.

Facebook updated their app constantly. They wanted people to stay connected all the time. Later, we would find that Facebook was using their app to spy on users, collect data, audio, location information and other statistics, then target us with personalized ads. Instagram had launched a year earlier, in 2010, and was becoming more popular every day. It was one of the first services whose functionality in the app version was better than it was on the web. Within two years, Facebook bought Instagram for a billion dollars. Snapchat appeared in 2011. I didn't use it, but it also began to explode, especially with younger people. Some people were becoming rich with the app economy and any business that wanted to remain relevant had to build their own app.

63

HURRICANE SANDY

In 2012, Hurricane Sandy hit New York City. My neighborhood suffered a power outage that lasted over five days. My world of technology took a tremendous hit. I walked to the Hudson River to see the storm surge and was horrified at the sight of water breaching the river wall and spreading over the sidewalk. Con-Ed, the power company, made robo-calls stating they might have to turn the power off. Two hours later, the lights went out and everything got quiet. I had frozen some gallon Ziploc bags with water, creating blocks of ice. I left these in the refrigerator and tried to keep the door closed as much as possible. My stove used natural gas, so I was able to cook, and my water arrived via pressure, not an electric pump. I had fire and water. Beyond that, everything else disappeared. I lit candles. As a closet survivalist, I had plenty of candles and matches, as well as Bic disposable lighters.

My landline worked, but I had no one to call, as most of my friends used cell phones and most of the cell towers had no electricity. I turned my cellphone off to save power, then went to bed. When I woke up and took the Elko for a walk, I found a weak cell signal ten blocks south. I went to my van, in the parking lot at Pier 40 on the Hudson River, to assess the damage. The air smelled of sewage and diesel fuel. My van was above the storm surge and I began to use it as a mobile command post. I went there every day to charge my phone, which I used as a hotspot for my laptop. I would check email, monitor the news, and post updates to Facebook, which functioned as a virtual community now. At home, I had my candles, my transistor radio, the landline, and the gas stove. These twentieth-century technologies were the only ones that still worked.

I was able to function, but in a diminished way. I used the last of what was in my fridge, took cold showers, barely went online, and talked little by phone. Nights were dark and scary, with no street lighting. On the second morning, strangers with Honda 2000 portable generators set up shop on street corners and sold hot coffee. I called them the "generator coffee guys,"

People were coming downtown with flashlights. They were blackout tourists who wanted to see what was going on. The electricity was working above 30th Street, less than a mile north, and I could see Battery Park City lit up. By the second night, some of my neighbors said they were "going uptown," leaving the blackout area. It felt like they were abandoning ship. They were going to stay with friends who had electricity, but I saw it as a betrayal to the neighborhood and decided to gut it out. By the third evening, though, I wanted pizza, so I got on my bicycle and went uptown. Life was going on as before and I felt a shock, like someone who had crossed over the border from East to West Berlin. I bought some non-perishable orange juice, got my pizza, and pedaled home.

I sat in the dark and thought about how alternating current was one of mankind's greatest inventions. I thought about all the things I could no longer do, such as print out my photos, read comfortably at night, and run air conditioning. Still, I had a focus during that blackout that I lacked when surrounded by my usual distractions. All I had to do each day was to eat, dress, walk the dog, and get to the van. Power came back, first to the Lower East Side, then to Chelsea. After 5 days, my brief plunge into a pre-electrical society ended. Whatever clarity I got from the blackout soon disappeared. I soon forgot about the whole experience and resumed my twentieth-first-century life.

64

MODERN LIFE

LOGGING INTO MY PHONE USED TO MEAN ENTERING A passcode, but fingerprint ID arrived in 2013 and I began to use my thumb to unlock the phone. It did not always work perfectly, especially in the cold weather, when my thumbprint seemed to get muddy. It saved time when it did work, though. "Biometrics" is another way of saying "body measurements," from fingerprints to cameras and software that precisely record, measure and store your virtual face. Handprint and fingerprint biometrics were being used at the NYU Medical Centers by this point. They made me place my palm in a box-like device before they would admit me for a routine test in 2013. It was to make my experience better, they told me.

The Department of Homeland Security used biometrics even earlier, for the Global Entry program. I had been fingerprinted in the 1980s and had always assumed that my prints lived somewhere in a permanent database. When the new iPhone requested a fingerprint scan, I submitted without a fight. It was not a big mental step from scanning a bar code at the supermarket to scanning my thumb to unlock the phone. It's a tremendous leap, though, to use technology for scanning, identifying and indexing people in public places with security cameras. When and where can we now expect privacy?

By the second decade of the twenty-first century, I was not just living more of my life "online." The internet had become part of my life in ways I never would have imagined. It was a progression I unaware of until I had gone far down that road. At first, every new function was a triumph of convenience. It started with a book that I bought on Amazon. I was soon buying other things online, not just books, but anything I could get cheaply and delivered easily. I bought clothes, camping gear, camera equipment, soap, sneakers, toiletries, dog toys, vitamins, food, computer components, power tools, musical equipment, automotive parts and more. It felt as though shopping online saved me time. I could make a few clicks and save myself another errand.

I paid my bills online and began to accumulate so many passwords that I thought about buying a password manager app. I didn't want one for philosophical reasons, but I thought about it. Some stores and services now required online payments. They did not want to process paper. For the ones that still took paper, I could send a paper check from my bank's website. I could go for years without visiting a physical bank. I read the newspaper online. I watched videos of animals doing silly things. I answered my email first thing in the morning, all day, and just before I went to bed.

I placed and answered ads for musical gigs, "For Sale" items and other things. I looked for work on specialized websites. I looked up almost anything. To "google" became a verb and the word "research" would never again mean the same thing as it did before. Everybody was suddenly a researcher. If a thought or a word was on the tip of my tongue, I could whip out my phone and address the collective memory of the internet to find an answer.

"Hold on a second," I would tell a friend, and they would understand that I was looking something up. I could visit online specialist forums and read the latest news and rumors. I could read about problems, raise questions, get answers, help people out, and get into arguments over stupid topics. I could compulsively check the weather all day. With all the time that the internet was saving me, I could devote hours to doing other things on the internet.

All my entertainment needs were met online. Why go out and buy an LP, a CD, or a DVD when I could stream almost anything from a service like Netflix, Hulu, Spotify, Pandora or YouTube? What I'm giving up in tactile sensations and the thrill of acquisition, I'm making up for with less clutter. I am now grateful I sold my records for drugs in the 1980s. I would have made more money had I kept them and sold them now, but I would also be living in a record closet.

I bought the first edition of the Apple Watch when it was released in 2015. I missed my Casio watches, though, so I sold it on a website that lets you sell old electronic gear. By the third iteration, in 2017, they had added cellular capability and made it waterproof enough to swim with. I tried it again and use it more often. It keeps track of my exercise and heart rate. I get sick of the watch's constant exhortations, "Breath" being one of them, along with "Time to Stand" and "Keep it Going," but I'm too lazy change the settings. I could do it if I wanted to devote some more time to customizing my technology.

2017 is when "Face ID" – facial biometrics – came to the iPhone. When this happened, some people worried about the facial information being stored, but most just used the technology. We have again surrendered to biometrics without a struggle.

65

THE PAST

I AM MORE NOSTALGIC NOW FOR OLD TECHNOLOGY THAN I would have guessed possible. My excitement for the latest and greatest technology has dimmed. I used to hate the term "digital immigrant," which is the opposite of a "digital native." I took umbrage at the idea that we who pioneered this modern, digital, interconnected world were not considered "natives," but that was my ego. "Digital immigrant" just means we were not born into the wired universe. We know what the world was like before this all came along. We are prelapsarian. We remember. I am reminded every day of an irretrievable past.

Phone booths are remnants of earlier times. I love to see them now. You can still find a few, usually built into the wall of an older building or a public place. I remember stepping into a booth and closing the door, getting that slight hush from the noise of the city. If you used a phone booth in a building or a hotel, there might be a stool and a small steel or Formica counter, along with a phone book that swung out from under the counter. In older booths, a light and a fan would come on automatically when the door folded shut.

The phone booth was an intimate space. Before cell phones, as late the year 2000, this was how you called people from the street. Much drama played out at phone booths. Drama now plays out on the street and we all have to listen to it, from people on their cellphones. The majority of people now walk down the streets talking, texting, or holding their phones in front of them while video chatting.

Making phone calls is not as popular as it once was. People text or use other apps. Most people consider public phones to be filthy and unusable. They are only good as a platform for advertising. Using a payphone has become a sign of class and status, just like taking the bus in Las Vegas. The only people who take public transportation in many large American cities do so because they can't afford a car. Using a payphone signals desperation in the twenty-first century.

Browsing in bookstores and record stores has become another dying pastime. Bookstores were once a destination. You could daydream, read magazines, meet other people, and add to your book or record collection. You would be in the physical presence of like-minded souls and other readers. In a record store, you could glance over and see what other people were buying. Who was that person going through the Lucinda Williams CDs? Who was at the "Self-Help" section in the bookstore? Bookstores still exist and they are still oases, but they're an endangered species. The internet has killed many of them.

I loved buying the Sunday Times, something I did every Saturday night for decades. I would wait outside a delicatessen for the newspaper truck to drop off the papers that just came off the presses. They were assembled with other sections by one or two workers, who would create a small mountain of Sunday papers. This continued into the 2000s. People would check to make sure that all the sections were there. You can still get the Sunday Times, but the local delis might have a pile of 40 or 50 copies now, instead of the mountains they had in the past.

People now read the Times, or whatever their local paper is, on their phone. Reading a newspaper has become far less common and reading it on newsprint has become a statement. It's not the same experience you get reading it online. From the smell of the newsprint to the way we scan the page for news, reading an actual newspaper is radically different. It's tactile and sensory. Pages crinkle, ink may smudge, the text may rip. You can doodle on it while you talk on the phone. You can rip out an article, fold it up and carry it around. You can use it to wash a window when you're done or crumple it up and use it for packing material in a box.

I used to pay for everything with cash, even though I had a credit card. I don't like to incur debt. I began to notice people using credit cards for small purchases after the year 2000. Many stores resisted. Card transactions cost them extra money, but the market demanded it and we are on the way to a cash-less society. When Apple Pay arrived, in 2014, it took a while for stores to accept it, but touch payments began to catch on.

I still carry cash, but I don't use it as much as I used to. I marvel at people who carry no cash. Some places in my neighborhood no longer accept cash, even though it's legal tender. Machines in supermarkets that count change and redeem it for paper money have become scarce. People don't keep big jars of change around anymore.

From the 1960s through the 1980s, I would circle upcoming shows in "TV Guide" or the TV section of the Sunday Times. In the 1980s, videotape and VCR timers made radical changes to my viewing habits. Cable networks gave me more choices. Now, streaming services have made live television almost obsolete. I have owned versions of the Apple TV since the second version came out, in 2010, but I watch little TV now. I cut the cord in 2018 and live easily without cable television. As with cigarettes, it took a few attempts before I finally quit. I sold my DVD collection, but not before ripping most of them to a hard drive, just in case I end up living in the woods one day.

When I was a child, airports were gateways to the future and the idea of flying was magical to me, even though I had never been on a plane. The experience of flying has been changing since it began, from how we eat on board, to the elimination of smoking, to how you plan and pay for a flight. We went from a simple paper ticket and boarding pass to electronic tickets. Then came boarding passes that we store on our phone. Security has gotten more complex, as well. In addition to the physical screenings and x-rays, we now submit to body scans.

We used to sit on flights and watch films that were projected from a reel onto a screen. Projectors were replaced by small TVs that tilted down from the ceiling along the aisle. Everybody on the plane had to watch the same thing. Personal screens began to appear on the seatback in front of us and we now have a range of movies to choose from. More and more people now watch content on their tablets and laptops. They bring their own entertainment, pre-loaded, to watch in their own bubbles. This is progress.

When you sat on a plane in the past, you were out of touch with the world. I saw my first GTE "Airfone" in the 1990s. It was in the seatback ahead of me and, despite the cost, I made a call to Hertz in San Francisco. I wanted to make sure the car I reserved had a cassette deck in it. I asked them the size of the engine displacement. They must have thought I was crazy. What I really wanted, though, was to speak with someone on the ground while I was in the air. That call cost me ten dollars. Planes now offer WiFi.

Technology steamrolls the past, faster and faster every year. A good friend, a successful corporate lawyer, told me of his love and loyalty for his Blackberry: "Within a week, I was addicted." After using it for a while, his feelings were "you'll have to pry it out of my cold, dead hands." When iPhones were new, most corporate IT departments would not support them and most companies would not pay for them. When the Blackberry began to fade, my friend would visit his IT department and request the Blackberry model he most loved from their dwindling stock. He was loyal to his device. It was only after his daughter opened his eyes to the iPhone that he finally gave up his beloved Blackberry.

I was incredibly loyal to my first email program, Eudora. After it was discontinued and became unstable, I moved to Thunderbird. For many reasons, I moved to Apple Mail and then back to Thunderbird again. Through all these switches, I held onto all of my old email, tens of thousands of emails, going back to 1995. I had to go through technical hoops to preserve all of them. They're important to me, but I am a sentimental geek and an archivist. I made the effort. I don't think most people would.

66

PROGRESS

When I was young, the idea of "progress" was an article of faith. We believed that science and technology would continue to advance and improve our lives. I believed this for a long time. My proof of this, as a child, was the space program. As I got older and learned about recent history, my belief did not waver. From the supreme effort to harness atomic energy to the antibiotics that cured my mom's tuberculosis, science solved problems. Black and white TV turned to color. Technology became more compact and available. There was palpable proof of progress in every corner of my life.

Over the course of three generations, from the birth of my grandparents until now, humans invented telephones and learned to speak with each other around the world. We conquered distance with airplanes and cars, learned how to record sound and images, extended life with medical technology and revolutionized farming with science. We created awful, but sophisticated, weapons of war, put a man on the moon, decoded the human genome, connected most of the planet through the internet and invented hand-held devices that allow for two-way video conversations with people around the world. The past forty years have seen more rapid technological change than any other era in recorded history.

The cost of progress may not have been as visible, though. We are now reaping the toxic results of the plastics revolution, the age of atomic waste, and the degradation of the environment through chemicals and industrial pollutants. We are the ones who live in a world that is getting hotter each year due to fossil fuels. We are confronting bacteria that's resistant to antibiotics. Our identity can be stolen by someone thousands of miles away. More objects than ever are disposable, from coffee cups to water bottles. This is all measurable - the physical consequences of our forward march. What, though, are the mental and spiritual results?

Human interactions have become disposable in many ways. Chats and messages can have little content and disappear instantly. More and more, we are each in our own bubble. We interact with each other less and barely live in the moment. In big cities, people walk immersed in their phones, heads down, missing the world around them. On the road, they drive while looking down and texting. People visit tourist spots to take selfies, even at memorials to tragedy. A person will re-post something on social media and consider it a concrete action. Here is what modern life has become – find an "experience" or location, photograph yourself and share it on social media.

Our language has changed, and not just with neologisms and new phrases for old words. People speak in shorter sentences. They hold conversations while looking at their phones and pay little attention to the person they are speaking with. When writing, it's common now to ignore punctuation and capitalization. Whole conversations take place with acronyms and emojis. Proper syntax has become technological roadkill. This is more than simply the current version of slang. It cuts across age, gender, class and ethnic boundaries.

Modern technology has changed how we communicate with each other. If I send an email or a text that's too long, people won't read it. Attention spans have gotten shorter. Social protocols have broken down. Many people are uncomfortable with telephone manners. There is often no civil way to end a text-conversation, aside from maybe a "thumbs-up" emoji or a "kk." It could simply be that etiquette needs to catch up with technology. Or it could be that the need for basic etiquette has vanished, that the world has gotten faster, harsher and colder.

What has this explosion of technology cost me? From 1995 to 2005, I did not walk around on many errands. I did not listen to the sounds of the city where I lived, I did not have many random interactions with neighbors and strangers. I was not getting lost in bookstores, sitting in the aisles, reading magazines, the way I had until the late 1990s.

One of the best antidotes to losing myself online was the dog I found on Interstate 80, in Nevada, in 2006. It's impossible to ignore a dog's needs. We would take long walks and I found myself enjoying the world again. It was a joy to simply walk and take in the world around me. I had not considered that my online activity was depriving me of experience in the real world. I began to question whether I wanted to cede so much of my time and privacy to the Gods of Silicon Valley.

People first talked about "internet addiction" in the 1990s. The bar for that ailment has gotten higher now. You have to be an obsessive gamer, someone who sits in an internet café for 36 hours straight, without eating, to be considered an internet addict. That view may be changing, though.

As of 2020, just 13 years after the introduction of the smart phone, journalists have begun to tell us how to cut down on "screen time." Apple will monitor it and measure it for you and inform you how much time you spend in the two-dimensional world. Children of Silicon Valley executives, we are told, are kept away from screens at a young age, as are their nannies. Most everyone else in the population, though, seems to be completely immersed in the virtual world of the internet, social media, and various apps.

We have become slaves to our passwords and need apps, which also need a password, to secure our passwords. The price for being lazy is high. An individual or entity can steal our identity and commit fraud in our name. When we forget passwords, we are conditioned to just groan and follow the drill of resetting them. Due to security requirements, I have to "prove" that "I'm not a robot" a few times every day.

I walk the streets and dodge adults on electric scooters who are speeding and wearing full headphones. I have to watch out for distracted people on bike shares and skateboards. They ride with earphones and AirPods, staring into their phones as they roll by. I listen to one side of many banal public phone conversations every day. I go through a finger-swiping dance at the checkout line while paying for something. The purchase gets recorded, a receipt comes to my email, and another data point on me gets inscribed into the Book of Life. This may all be used against me one day.

Everything I have ever posted online will probably persist. Records that I thought were private will exist whether I want them to or not. My life involves regular, ongoing battles with tech support. Sometimes, I'm in the middle of an automated phone menu and a sneeze will set off the next prompt. I'll get shunted from one support person to the next and have to explain the same issue multiple times. Automated menus are an abomination of modern existence. Few things are so dehumanizing as a robot that pretends it understands you. The main purpose of a phone menu or robot is to prevent you from speaking with a human, which incurs more cost for the monolith that you're calling.

I have had to learn to be patient. The passwords, robots, fraud, spam, and distracted fellow humans all seem like necessary evils. They are the price I pay to stay connected to life and people, but I sometimes question whether these evils are worth it. At some point in the late 1990s, when the digital waters had begun to rise around me, I began to have nostalgic fantasies for a life that I never experienced. I fantasized about living in a cinderblock bunker, way out in the desert, with only a landline and a typewriter. These fantasies came from a desperate need to escape from modern digital life. I have generally left this need unfulfilled, but it has grown more pressing.

Our new, wired world was created by the convergence of three technologies: the internet, cellular networks, and smartphones. Each of these technologies arose on its own. There was no grand plan to create a device that was portable and networked with billions of other devices and nodes. No author or visionary, even Vannevar Bush, dreamed of the unintended byproducts from the technology we live with now. Our technology defines us more and more, through the devices and services we use and those we don't use.

I have no answers. I leave any conclusions to others, but I do know that predictions for the future are almost always wrong when that future rolls around. We know that technology has costs – now and in the future. We know that technology can solve problems. Technology and the world made by humans will not disappear. It's unavoidable. A life mediated by the internet will not disappear, either. In many ways, the internet has been the biggest revolution seen by humankind. It's too complex, too close and we are too much in the midst of this change to see the faults and benefits clearly now. Things are still fluid.

Will this pace continue? Who will make the decisions about such things as the ethics of "robots versus humans," about gene splicing and other conflicts that science is producing? Will the legal mechanisms catch up to the pace of technology? Truth and facts have become "alternative" and are up for debate. When you cannot agree on facts, how do you confront them? And what about the rest of the globe?

We are more interconnected each day. Despite the inevitable, cyclical turns inward, towards toxic nationalism and tribalism, there is no turning back on globalism. Countries may turn off their internet infrastructure when they have unrest, but they cannot completely stop the flow of information. Everything is being documented and shared now, for better or for worse. We are still in the wave and we don't know how big that wave is.

Decades after I moved away from Iroquois Road, the dead-end street where I grew up and the launching pad for my early space travels with Roland, I returned to see what it looked like. The trees were all gone. The people did not want to rake leaves any more. There were cars up and down the street. Some people had paved over their lawns so they could park more cars. The houses all seemed smaller to me. No wonder - I was bigger and I had also seen more of the world. The houses were physically different. They did not match my memory. They had been modified, enlarged, painted and modernized. The street seemed narrower, even though that was not possible. As a child, that block had been my entire universe. The huge piece of granite, the rock, was still in the dead end, though. I went to visit P.S. 28 and it was run-down. The dirt field remained, but I was not even sure the school was still open.

The apartment buildings where I lived in the 1980s still stand and still look the same from the outside. The building in Chelsea has looked the same since it was built in the 1870s. The building on Barrow Street is unchanged on the outside, but the hallways inside have gotten more modern. Most of the facades in Greenwich Village, though, are protected. Something else has changed. The stores are no longer mom and pop operations. They are either empty storefronts or flagship operations for global brands. The people have changed the most. They don't sit out on stoops anymore, watching the world go by. Everybody is in more of a hurry and they are lost in their electronic worlds.

I have lived at my current place since 2002. It's filled with more stuff now than when I moved in. Stasis will do that. I have walls of books and piles of camera equipment. The shelves of CDs and DVDs are gone, but everything else looks the same. Santo, the young dog whom I found with the help of Twitter, rules the roost, usually from his couch. Road atlases are scattered all around the house. The workstation I built as a temporary expedient when I moved in is still there, covered with gear. The computers have gotten faster, the backup drives have gotten bigger and the connection speed has gotten faster. I still have the van, but it's getting old and I may sell it. I may even move someplace quieter.

I have a small box of things that belonged to my dad, a notebook from my mom, and lots of family photos. I look at the scans once in a while, especially if I want to feel sad. I have a television, but never watch it. Somewhere in a shoulder bag at the back of a closet are dozens of love letters I received from a girlfriend in the early 1980s. I have not wanted to read them for decades. I may never, but I know where they are.